T0332039

Introduction to

Nanoelectronic
Single-Electron Circuit Design

Introduction to
Nanoelectronic
Single-Electron Circuit Design

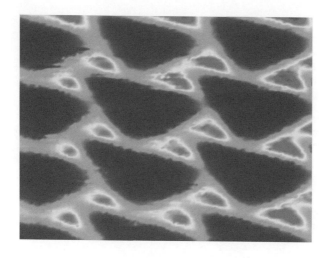

Jaap Hoekstra

Delft University of Technology, The Netherlands

PAN STANFORD PUBLISHING

Published by

Pan Stanford Publishing Pte. Ltd.
Penthouse Level, Suntec Tower 3
8 Temasek Boulevard
Singapore 038988

Email: editorial@panstanford.com
Web: www.panstanford.com

British Library Cataloguing-in-Publication Data
A catalogue record for this book is available from the British Library.

ISBN-13 978-981-4241-93-9
ISBN-10 981-4241-93-8

Printed in Singapore by B & JO Enterprise

Preface

In *Introduction to Nanoelectronic Single-Electron Circuit Design* single-electron circuits are studied as an introduction to the fast expanding field of nanoelectronics. In nanoelectronics, single-electron circuits are those circuits that process information and signals by making use of time-dependent currents and voltages due to charge transport by just one single or only a few electrons.

This textbook follows an unconventional approach to explaining the operation and design of single-electron circuits. In general, the conventional approach to this subject is to begin with a brief introduction to the quantum physics of the nanodevices followed by modeling the devices by mathematical means. It is the author's opinion that an alternative approach with an emphasis on experiments to obtain a characterization of the devices will enhance the reader's comprehension. Therefore after the introduction, first some landmark experiments are reviewed and discussed. Then a brief introduction to the relevant (quantum) physics of the nanodevices is given. Subsequently, the characterization of devices is used to obtain equivalent circuit diagrams. To ease the discussions on the characteristics and the equivalent circuits some topics from linear and nonlinear circuit theory are briefly reviewed. Knowing this, a circuit theoretical framework will be built. Devices and (small) circuits are modeled both by mathematical means and by circuit simulations. Also currents in classical and quantum physics are reviewed. Simple circuits including single-electron devices are treated. After this, circuit design methodologies are discussed as well as typical electronics' topics as signal amplification, biasing, coupling, noise, and circuit simulation. When looking forward to dealing with systems emphasis is placed on redundant and fault-tolerant architectures to cope with the uncertainties related to the critical nanometer-sized device dimensions and the "probabilistic" nature of quantum physics. The book ends with a brief discussion on potential applications and challenges.

Due to their simplicity, this monograph mainly considers *metallic* single-electron tunneling junctions (metallic SET junctions). For an introduction to single-electron circuit design, circuits with these devices are already complex enough.

The basic physical phenomena under consideration are the quantum mechanical tunneling of electrons through a small insulating gap between two metal leads, the Coulomb blockade, and the associated phenomenon of Coulomb oscillations—the last two resulting from the quantization of charge. The metal-insulator-metal structure through which the electrons may tunnel is called a tunnel(ing) junction. This tunneling is considered to be stochastic, that is, successive tunneling events across a tunnel junction are uncorrelated, and is described by a Poisson process. Throughout the text tunneling through a potential barrier is considered to be non-dissipative (the tunneling process through the barrier is considered to be elastic), unless explicitly stated differently.

Electron transport in the nanoelectronic devices can best be described by quantum physics; nanoelectronic circuits can best be described by Kirchhoff's voltage and current laws, which have a firm basis in classical physics. This tension between quantum physics and classical physics is taken for granted; experiments with circuits will have to approve whether we can successfully include the quantum character of charge in a circuit theory for single-electron electronics.

As quantum mechanics is described in terms of energy, it seems obvious to describe, that is to *analyze*, the behavior of SET devices and SET circuits with energies. This is what the so-called orthodox theory of single electronics does. In this semiclassical-physics theory an electron will tunnel if the free (electrostatic) energy in the system after tunneling is lower than the free energy in the system before tunneling. However, as we will see, this energy loss cannot always be modeled as dissipation by heat or by radiation without violating the Kirchhoff laws. Especially, when the tunnel event of single electrons is considered the dissipation cannot be modeled by a finite resistance. To design, that is to *synthesize* circuits with these devices, we need a circuit theory. It must be based on Kirchhoff's voltage and current laws. In contrast, these laws ensure energy conservation in circuits: any energy dissipation in the circuit is delivered by sources; and *vice versa*, all energy delivered by sources is either stored either dissipated in circuit elements. It is because of these arguments that the orthodox theory of single electronics is not followed for designing circuits.

This text provides a circuit theoretical model of the single-electron tunneling junction to analyze and synthesize nanoelectronic circuits. In the absence of tunneling, the metallic junction is modeled as a capacitor. Tunneling is explained by considering the (matter) wave nature of the electron, expressed by the De Broglie wave length $\lambda_F = h/\sqrt{2mE_F}$. Two metallic junctions in series form an island. To model the two junctions as capacitors (for example, in case of predicting Coulomb blockade) the island must be large compared with λ_F; that is, unless explicitly stated quantum-dot systems are not considered. The main results of this approach are that instead of attributing the blockade phenomenon to the existence of islands, the Coulomb blockade is found as a property of the tunnel junction for non quantum-dot systems, and tunneling must be modeled by an impulsive current source.

In chapter 1 nanoelectronics and single-electron electronics are defined and the scope of this text is presented. A bird's eye view is presented, with many pointers to later chapters, in order to familiarize the reader with the kind of possibilities and challenges this book is dealing with. Chapter 2 discusses landmark experiments that form a chain from the first experiments showing quantum mechanical tunneling to experiments showing Coulomb blockade, that is, no tunneling is observed while tunnel events were expected to happen. Chapter 3 consists of a brief review of the modeling of currents in classical physics and introduces circuit theory using lumped circuit elements, the chapter is essential for understanding of what can be or what cannot be modeled in a (electronic) circuit theory. Chapter 4 focusses on the quantum mechanical description of free electrons. In quantum mechanics electrons are described both as particles and as waves. Typical quantum mechanical phenomena as energy quantization and tunneling are possible due the wave nature of the electron. This chapter is the first of two that introduce the quantum mechanics needed for the understanding of the nanoelectronic devices. The second chapter, chapter 5 treats the quantum mechanical descriptions of currents in general, and the tunnel current in particular. Ballistic transport and quantized resistance are presented briefly. Chapter 6 the relation between lumped circuits, Kirchhoff's laws and energy in circuit theory is examined. It focusses on the conservation of energy based on Tellegen's theorem. The concepts bounded and unbounded currents are introduced in chapter 7, where energy conservation in the switched two-capacitor circuit is discussed. Also the initial charge models for the capacitor are presented. Knowing the circuit theoretical basics of capacitor circuits, the impulse circuit model for single-electron tunneling is presented in two chapters. First in chapter 8, based on energy conservation and a hot-electron model, the impulse circuit model is derived in case of zero-tunneling time. In chapter 9 the model is extended to nonzero tunneling times, to circuit including resistors, and to circuits excited by nonideal energy sources. Also, tunneling of many electrons in the same time interval is considered leading to the definition of the tunnel resistance. In chapter 10 the theory is generalized to multi-junction circuits. Especially, answers are discussed to the following question: How much energy is needed to tunnel onto a metallic island? Chapter 11 applies the impulse circuit model to the most basic single-electron tunneling circuits. It treats the electron-box, SET transistor, three-junction structures, and the SET inverter. Knowing how to analyze single-electron tunneling circuits chapter 12 starts the discussion on nanoelectronic circuit design methodologies and SET circuit design issues. As examples of possible circuit solutions to coping with uncertainties and inaccuracies SET based artificial neural network building blocks are described. The last regular chapter, chapter 13 gives an outlook to potential applications and challenges. At the end of this book an epilogue is added, especially for those readers who are already familiar to the orthodox theory of single electronics for circuit design.

I must thank all my colleagues, Ph.D. and master students that in many ways contributed to this book. Especially, I wish to thank Martijn Goossens, Chris Verhoeven, Jose Camargo da Costa, and Arthur van Roermund for introducing me to the field of nanoelectronics, and for the many discussions on electronic design methodologies; Roelof Klunder for the discussions on the SET circuit design issues and the development of the electron-box logic; and Rudie van de Haar for developing the Spice simulation environment, and his design of the neural node. For the more general discussions on nanoelectronic architectural issue I thank Eelco Rouw. Their Ph.D. theses can be downloaded from the library site of the Delft University of Technology. My special thanks must go to my wife, Judy, and my sons Tom, Jeroen, and Peter. They always encouraged me to write this book.

The textbook is based on a nanoelectronics course at the Delft University of Technology and is intended for senior undergraduate and graduate students. The prerequisites for understanding the material are the basic principles of solid-state and semiconductor physics and devices, and the basic principles of linear circuit analysis.

Jaap Hoekstra

Contents

Chapter 1

Introduction

In this chapter the topic of nanoelectronics is introduced and a bird's eye view is presented to familiarize the reader with the kind of possibilities and challenges this book is dealing with. Successively—the general scope, electron tunneling, tunneling capacitors, island charges, bounded and unbounded currents, and energy in simple capacitor circuits are introduced briefly.

1.1 Scope

Due to the ongoing downsizing of microelectronic circuit components, many nanoelectronic devices have been proposed and manufactured in the past years. These **nanoelectronic devices** have critical dimensions of several nanometers and take advantage of quantum mechanical phenomena that appear at nanometer scale.

Using these criteria, the tunnel diode or Esaki diode (section 2.1) can be seen as the first nanoelectronic device. Its operation differs from the injection of electrons and holes across the potential barrier in ordinary $p - n$ junctions in that electrons also tunnel through the potential barrier, because the barrier of the tunnel diode is extremely narrow.

Another nanoelectronic device is formed when the capacitor is scaled down. If the insulating layer is thin enough, again, tunneling of electrons through the dielectric from the negative to the positive side may occur (section 5.3). Consequently, not only a displacement current but also a conductive tunnel current can flow through the capacitor. Combining both currents, the nanoelectronic capacitor may be described as a tunneling diode parallel to a normal capacitor in first instance. The model is a suitable circuit model when we consider currents; the modeling in this case is sometimes called macro-modeling.

For individual electrons, however, the model is not correct. Due to the quantization of charge, tunneling of single electrons can be observed. Chapter 2 deals with experiments showing the effects of tunneling of single electrons: Coulomb blockade and Coulomb oscillations. To include the possible tunneling of single electrons the model of the capacitor and the parallel tunnel diode cannot be used; instead of this a circuit model based on the quantization of charge is developed.

Because of the capacitor part in the equivalent circuit models, basic capacitor circuits are discussed throughout the text. Special attention is paid to the energy balance in those circuits. It will turn out that this issue is of great importance to come to both the equivalent circuit of the tunneling capacitor and the condition for tunneling of single electrons to take place.

1.1.1 *Nanoelectronic circuit design issues*

To understand the possible utilization of nanoelectronic devices, useful and competitive circuits have to be designed. New circuit ideas must be developed exploiting the quantum character, the small feature size, and the low power operation of the new nanoelectronic devices.

There are a couple of different definitions of nanoelectronics around today. From the physics point of view, **nanoelectronics** often deal with small circuits including nanoelectronic devices in which dimensions have reached such a small length that the wave nature of the electrons cannot be neglected, and that device- and circuit simulations for basically classical device structures are confronted with a full quantum mechanical description rather than with mixed quantum-classical models. From the electrical engineering point of view, **nanoelectronics** is understood merely as an electronics based on nanoelectronic devices which utilize quantum mechanical phenomena; they have to be described with quantum-classical electrical models [Csurgay (2007)] to make circuit synthesis possible. This text follows this last point of view.

As a first introduction to nanoelectronics, **single-electron circuit design** and **single-electron electronics** are studied; that is, nanoelectronic circuits are studied in which information and signal processing are considered by means of transport of one single or only a few electrons. Especially, as signals that carry the information are considered: the value of a voltage as a function of time, the value of a current as a function of time, or the number of additional electrons on isolated islands as function of time.

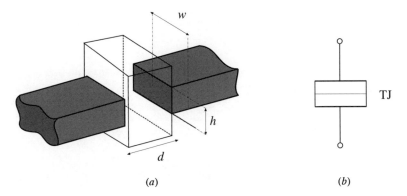

Fig. 1.1 (a) Metallic single-electron tunneling (SET) junction schematically: two metal leads separated by a thin layer of insulating material; (b) symbol for the SET junction circuit element

As this text is focused on circuit design, the nanoelectronic devices are introduced by measurements of their characteristics. After this the modeling of the components both by mathematical means and by circuit simulations is treated; circuit-design issues are discussed. For this purpose, first, equivalent circuits are proposed. After that, simple circuits, systems, and typical electronics' topics are treated (section 12.3).

To explain the important aspects of single-electron electronics, circuit design with **metallic single-electron tunneling junctions** (metallic SET junctions) is examined. Schematically, the metallic tunnel junction consists of two metal conductors separated by a very thin insulator (typically: 3 nm $< d <$ 10 nm), see figure Fig. 1.1. Due to the extreme small insulator thickness quantum mechanical tunneling of electrons through this insulator becomes possible.

1.1.2 *Levels in modeling, a top-down approach*

To follow the organization of the book, take a look at figure Fig. 1.2. It shows different levels in modeling that can be distinguished when treating the various topics involved in nanoelectronic circuit design.

It is necessary to understand that any modeling effort of a real device involves a theoretical framework on a certain level. Modeling at the upper level is used for the design at the system level (chapter 10); it takes complete sub-circuits (e.g., a full adder) as its inputs. One level below we model (sub)circuits; it is called the circuit theory level due to the fact that circuits

DIFFERENT LEVELS IN MODELING   MODELING EXAMPLE:

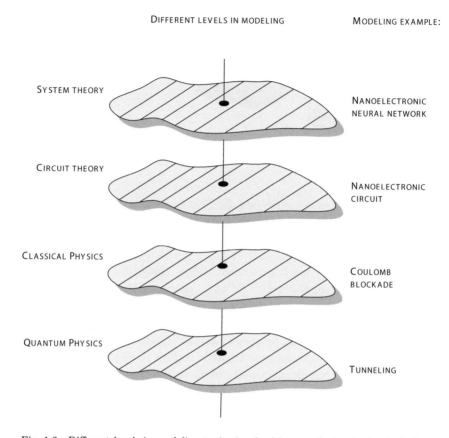

SYSTEM THEORY

NANOELECTRONIC
NEURAL NETWORK

CIRCUIT THEORY

NANOELECTRONIC
CIRCUIT

CLASSICAL PHYSICS

COULOMB
BLOCKADE

QUANTUM PHYSICS

TUNNELING

Fig. 1.2   Different levels in modeling topics involved in nanoelectronic circuit design.

are designed according to the Kirchhoff laws (chapter 6). This level is an approximation of the classical-physics level that takes the full Maxwell equations (chapter 3) into account. It is important to distinguish between these two levels; in this text the devices and circuits are sometimes analyzed on both levels. On the classical physics level, devices and simple circuits are analyzed with energy-band diagrams; and later, on the circuit theory level the devices and circuits are analyzed with the impulse equivalent circuit model. Finally, on the lowest level devices are analyzed on the quantum-physics level (chapters 4 and 5).

Sometimes, the same device or device property can be modeled at different levels. For example, a transistor can be fully described by semiclassical physics on the classical-physics level; but for circuit design we often can suffice with a description of the transistor's characteristics (e.g., in the *vi*-plane) on the circuit-theoretical level. In other cases, a typical phenomenon can be modeled correctly at only one level, for example, the tunneling of electrons can only be described in quantum physics. If we want to model this phenomenon at another level, the best we can do is to find an approximation that is suitable.

The important message in figure Fig. 1.2 is that the various levels really differ in the way they describe the physics and in the way they analyze circuits. They have different assumptions and use different tools. For example, the quantum-physics level differs from the classical-physics level by considering the electron not only as a particle but also as a wave, the circuit-theory level differs from the classical-physics level because it does not include the release of electromagnetic radiation in circuits in a general way, and the circuit-theory level differs from the system-theory level by taking much more details into consideration and by using different simulation tools.

1.1.3 *Overview of tunneling capacitor circuit models*

The equivalent model for the tunnel junction presented in this monograph is not the first attempt to model the tunneling capacitor. Since the measurements of tunneling between two parallel metallic plates by Giaever in 1959, various models have been proposed. A short overview is presented, it shows the development of circuit ideas to obtain an equivalent circuit for the tunnel(ing) capacitor. Notice that even for probably the most simple nanoelectronic device, such as the tunnel(ing) metal-insulator-metal junction, various proposals already exist. Obviously, even the simplest tunnel device is not so simple to model.

Historically, a good starting point is the modeling of two metals with a larger than nanometer separation by a capacitor—see figure Fig. 1.3 (0). This is, of course, how electronics started two centuries ago. The discovery of the tunnel effect in the tunnel diode [Esaki (1958)] (section 2.1) and the measurement of tunneling between two metal plates [Giaever (1969)] (section 2.2), both in 1958-1959, led to a model in which the tunneling junction was represented by a "tunnel resistance". This was a reasonable model, because the *vi*-characteristic, of the tunneling plates excited by a voltage source, showed a linear relation for small voltages, see figure Fig. 1.3 (1).

Based on research of Giaever and Zeller on small metal particles embed-

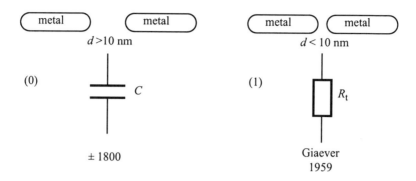

Fig. 1.3   Proposed equivalent circuits for the metallic SET junction - I; d is the separation between the two metals.

ded in an oxide film [Giaever and Zeller (1968)] (section 2.2), and of their own on research on what we would now call a single-electron box (section 2.2), Lambe and Jaklevic proposed the parallel capacitor/resistance model [Lambe and Jaklevic (1969)] of figure Fig. 1.4 (2). In both studies, tunneling is represented by a resistor, and anomalies in the characteristics are explained by introducing a capacitor. The next important step had to be taken when Fulton and Dolan directly measured the Coulomb blockade phenomenon in 1987 [Fulton and Dolan (1987)] (section 2.2). The current through a metal-insulator-metal-insulator-metal (MIMIM) structure is blocked for very small values of the applied voltage (the circuit consisted of a MIMIM structure excited by a voltage source). For values of the source larger than a critical voltage the resistive behavior is shown again. The in this way obtained characteristic is an affine linear relation between the voltage across and the current through the device for voltages above a nonzero critical value (if the critical voltage is zero the relation will be linear instead of affine linear). To avoid dominating noise from the operational temperature, very small junctions and, often, low temperatures were (and still are) needed. The equivalent circuit they proposed was the parallel capacitor/current source combination of figure Fig. 1.4 (3). In fact, the current source they propose is voltage controlled, so in fact it is a nonlinear resistance. The idea of modeling a tunnel current by a current source is an interesting and, in fact, a natural idea.

While developing the orthodox theory of single-electronics, more or less at the same time, Averin and Likharev, again, proposed the description in which the tunneling junction was described by a parallel combination of

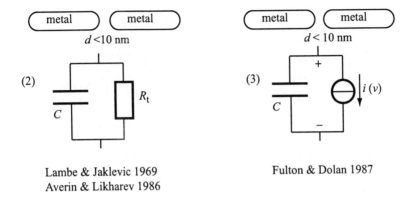

Fig. 1.4   Proposed equivalent circuits for the metallic SET junction - II

a capacitor [Averin and Likharev (1986); Likharev (1999)], if the junction is in blockade, and a resistor, if the junction is tunneling. The resistor is equal to the "tunneling resistance" R_t . The reason why the current source description disappeared was a consequence of the assumption that the orthodox theory neglects the tunneling time and the energy contribution of current sources.

Based on the measurements of the Coulomb blockade Devoret *et al.* proposed the a parallel combination of a capacitor and an ideal tunnel diode, shown in figure Fig. 1.5 (4) [Devoret *et al.* (1990)]. A model that is still very useful as a starting point for modeling currents through, for example, a single-electron tunneling transistor (macro-modeling).

Up to now, the parallel combination of a capacitor and either a linear resistor or a tunnel diode are the dominant equivalent circuits for the tunnel(ing) junction, see for example [Kasper and Paul (2005); Dragoman and Dragoman (2006); Hanson (2008)].

If we include, however, the predicted behavior of a single electron tunneling through a tunnel(ing) junction excited by a current source the parallel capacitor/nonlinear resistor combination does not correctly describe the predicted behavior. A solution is to describe the tunnel(ing) junction by the parallel combination of a (charged) capacitor and a (impulsive) current source as shown in figure Fig. 1.5 (5) [Klunder and Hoekstra (2001); Hoekstra (2001)]. In this model the tunnel event of a single electron is described by a impulsive current source of the form: $i = e\delta(t - t_0)$, in which $\delta(.)$ is the (Dirac) delta pulse, e is the elementary electron charge, and t_0 denotes the time the tunneling is modeled. The time t_0 is determined by, amongst

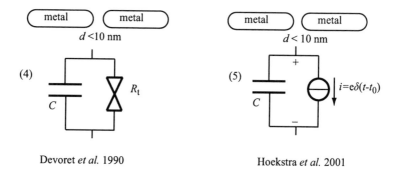

Fig. 1.5   Proposed equivalent circuits for the metallic SET junction - III

others, the so-called critical voltage. The model is called the impulse circuit model for the single-electron tunnel(ing) junction (chapter 7).

1.1.4   *Important quantum mechanical phenomena*

Within the context of this book, for nanoelectronics the following quantum mechanical properties of particles (here mainly electrons) and associated phenomena are important and will be treated:

- wave properties of electrons, which make it possible for them to tunnel through small potential barriers,
- the same wave properties, which cause energy quantization in quantum wells,
- energy quantization in a quantum wire (1-dimensional), which leads to quantization of resistance,

and last but not least,

- quantization of charge, which results in Coulomb blockade and Coulomb oscillations.

In particular, the above mentioned quantum mechanical phenomena lead to the following important applications using the quantization of charge:

- single electrons can be stored and retrieved in a so-called single-electron box,
- single-electron logic gates and other single-electron circuits,

and

- a description of the tunnel current based on quantization of charge and stochastic behavior in time.

1.2 Electron Tunneling

In quantum mechanics electrons are described by a wave equation for matter waves (chapter 4). It is according to this wave nature of particles that tunneling through a small barrier can occur. Electron tunneling deals with the penetration of a potential barrier, in case the total energy of an electron approaching the barrier is smaller than the potential energy height of the barrier. In this case, classical mechanics requires that the electron should be reflected back at the barrier. Quantum mechanics, however, allow penetration in or even transmission through the barrier, due to the wave properties of the electron. In case of transmission there is a finite possibility of finding the electron at the other side of the barrier, that is: the electron tunneled through the barrier. Tunneling through a potential barrier will be treated in chapter 5 in some detail.

For tunneling of electrons , a reservoir with free electrons will be needed. As we are going to consider electronic circuits a main reservoir is the metallic interconnect. Due to the conservation of charge, the sources—both current and voltage sources—only deliver the energy to circulate the current in the circuit.

1.2.1 *Free electrons*

In the metal many electrons are shared by all atoms and are free to wander through the crystal. These free electrons are hold within the wires by the edges of the metal, forming a barrier that confines the electrons. In this sense they behave like a gas, an "electron" gas (section 4.5). In terms of quantum mechanics the free electrons in the metal are considered to move freely in a potential well, that can be approximated by a rectangular potential well. As a consequence of such a description, the free electrons in the metal are actually stacked up in energy levels. The energy levels are very dense, since the well is, in general, very wide (section 4.5).

Because no more than two electrons can occupy any given level (Pauli exclusion principle), the lowest energy state of the metal shows a configuration where all levels up to the **Fermi level** are filled at $T = 0$ K. When the temperature is above 0 K, a few electrons are excited to higher levels.

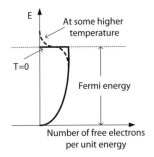

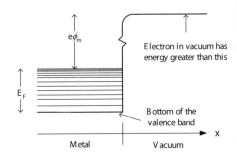

Fig. 1.6   Free electron distribution in metals.

Fig. 1.7   (Potential) energy relations at the metal-vacuum interface

The distribution of free electrons among the energy levels is shown in figure Fig. 1.6. For most metals holds a linear temperature dependence of resistivity. For example, pure Cu exhibits an almost linear temperature dependence in the temperature range from 80 K up to about 800 K.

The minimum energy necessary for an electron to escape into the vacuum from an initial energy at the Fermi level is defined as the **work function** $e\phi_m$ (ϕ_m in volt) as shown in figure Fig. 1.7. So, the electrons in the metal face a potential barrier at the metal-vacuum interface. For metals, $e\phi_m$ is of the order of a few electron-volt and varies roughly from 2 eV to 6 eV. The work functions of various metals are given in table 1.1.

1.2.2   *Tunneling*

Now, consider a simple tunneling system consisting of two metal plates separated only a few nanometers. Experiments show that most applied voltages across the junction are in the mV range (typically upto a few hundred mV), consequently, the conditions are such that electrons only have energies much lower than the energy necessary to free the electrons

Table 1.1   Work function of some metals

Metal	Work function, $e\phi_m$
Ag, silver	4.26 eV
Al, aluminum	4.28 eV
Au, gold	5.1 eV
Cr, chromium	4.5 eV
Cu, copper	4.65 eV
Nb, niobium	4.3 eV

Table 1.2 Fermi energies for free electrons in various metals

Metal	Fermi energy, E_F
Ag, silver	5.50 eV
Al, aluminum	11.65 eV
Au, gold	5.52 eV
Cu, copper	7.03 eV

from the metal. In fact, the kinetic energy of the tunneling electrons is between zero and a value just above the Fermi level; Table 1.2 shows some values of the Fermi energy for various metals.

In general, there will be a potential difference v across the junction. Suppose the energy levels of electrons moving on both sides of the barrier are as shown in figure 1.8 and the potential difference across both sides is considered constant before and after tunneling (an assumption that will be discussed later). If we use the same metal at both sides, the height of both metal-vacuum barriers are the same indicating that the energy to free an electron is the same at both sides, and electrons moving on the Fermi levels at either side of the junction have the same kinetic energy. The Fermi levels are ev apart.

1.2.3 *Hot electrons*

If we consider tunneling of an electron we notice that the total energy of the tunneling electron is the same at both sides of the junction because there are no external forces considered acting on the electron during the tunnel event. An electron tunneling from the left side towards the right side of figure Fig. 1.8 will therefore arrive at the right side on an empty energy level *above* the Fermi level as a *hot* electron. Apparently, when the tunneling electron is just in the metal on the right side, it has more energy than most other electrons at this side.

A **hot electron** is an electron with energy more than a few kT ($kT = 2.59 \times 10^{-2}$ eV) above the Fermi energy. The hot electron is not in thermal equilibrium with the lattice and the other electrons, and, in general this hot electron will lose its excess energy very fast due to collisions with the lattice and with other electrons.

For example, in figure Fig. 1.8, an electron leaving at the Fermi level on the left side will upon arriving at the right side be hotter by just an amount of ev compared to an electron on the Fermi level at this side. In case of

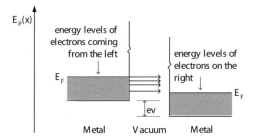

Fig. 1.8   Energy levels at the metal-vacuum-metal interface; the arrows represent possible tunnel events.

tunnel currents the hot-electron concentration drops exponentially towards the interior of the metal. One might even define an attenuation length in the direction perpendicular to the interface [Lindmayer and Wrigley (1965)]. The event of tunneling and subsequently traversing to the Fermi level will be modeled with a *tunnel-and-fall model* in chapters 7 and 8.

1.2.4   *Tunneling time and transition time*

The time associated with the passage of an electron under the tunneling barrier is called the **tunneling time**. The existence of the tunneling time is well accepted and has been measured experimentally to be in the order of 10^{-15} s. If the electron becomes a hot electron after tunneling, the time the electron needs to tunnel *and* to release energy while restoring equilibrium can be called the **transition time** τ.

One sees in figure Fig. 1.8 electrons can tunnel from all energy levels that face empty states. Such a picture is in agreement with experimental results on the tunnel diode and on the tunneling between metallic plates (chapter 2). In this example, all electrons on energy levels between E_F and $(E_F - ev)$ are able to tunnel at the metal-vacuum-metal interface.

1.3   Tunneling Capacitors and Island Charges

As our main concern is to model the working of the nanoelectronic devices according to a circuit theory, we also have to know on what energy levels the conduction electrons move in other parts of the circuit. The fundamental theory is here that current conduction is not possible in empty or completely filled levels. It is according to this notion that we can model the conduction electrons in the metal interconnect as moving in a small energy band around

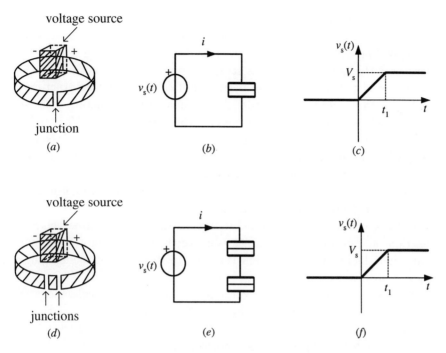

Fig. 1.9 (a) Example of a junction in a single free-electron reservoir; (d) example of two free-electron reservoirs created by two junctions; (b,e) circuit diagrams; (c,f) applied source waveforms.

or just on the Fermi level (section 5.1).

Going to circuits, two basically different structures can be distinguished in which tunneling can occur, namely: an electronically seen single reservoir filled with free electrons in which a small gap has been made, or multiple reservoirs separated by small gaps. The prototype for a single reservoir is a structure that consists of a metallic junction coupled to a voltage source, as shown in figure Fig. 1.9 (a), (b), and (c).

Particularly, the circuit is investigated at time t, $t > t_1$, when the voltage source has obtained a constant value V_s. In fact, this structure describes a solitary junction. In chapter 2, we will see that, the following experimental results have been obtained: (1) if the capacitance of the tunneling capacitor is large—for example in case of parallel plates—electrons will continuously tunnel if the gap is sufficiently small; (2) the capacitor will be charged first until a critical voltage is reached in case a substantial impedance is connected to it—that is, the energy source behaves as a current source; (3)

if the junction is very small, a resistance peak at zero bias voltage is found.

The prototype for a structure consisting of multiple reservoirs is shown in figure Fig. 1.9 (d), (e), and (f). Again in chapter 2, we will see experiments that show that in this structure tunneling may not be possible for values of the source voltage below a critical voltage; the structure is said to be in **Coulomb blockade**. An example of this structure is the so-called single-electron tunneling transistor (SET transistor). What makes this structure different from the previous one is the possibility of the existence of a non-zero reservoir charge; of course, due to charge neutrality the total charge of all reservoirs together is zero.

Measurements, in which we see tunneling in a metal-insulator-metal structure, indicate that the tunnel current in a SET junction circuit excited by a voltage source is proportional to the voltage across the SET junction (one uses sufficiently large values of the voltage source); this suggests that the metallic tunnel junction can, in this case, be modeled by a linear resistor.

However, other results of experiments show that tunneling can be blocked if a positive voltage across the tunnel junction is lower than a critical value, that is, we see Coulomb blockade. To explain the Coulomb blockade a junction capacitance is modeled. Let's investigate the structures in terms of capacitors and clarify what the influence of additional charge on a reservoir is.

1.3.1 *Two-junction circuit in Coulomb blockade*

If there isn't any tunneling, we assume the equivalent circuit of the two reservoirs of figure Fig. 1.9 (d) to be a series combination of two capacitors. If we allow the possibility that one or more electrons had been tunneled through only one of the junctions then the total net charge of the two reservoirs will be non-zero and opposite. In the equivalent circuit this means that the **floating node** between the capacitors can have a net charge. To emphasize this, the floating node is generally called an **island** and the net charge is called the **island charge** $q_i(t)$.

We will have to determine under what conditions the existence of such an island charge can be included in circuit analysis. In general, the requirements for including in a circuit theory, either linear or nonlinear, are that Kirchhoff's laws hold. So, we will proceed by finding an expression for the voltages across the capacitors that suffice Kirchhoff's voltage law and then consider the equality of the currents through the capacitors to fulfill the

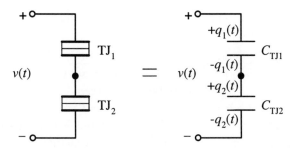

Fig. 1.10 Equivalent subcircuit for two tunnel junctions in Coulomb blockade.

current law, see figure Fig. 1.10. If $v_{\text{TJ1}}(t)$ is the voltage across C_{TJ1}, and $v_{\text{TJ2}}(t)$ is the voltage across C_{TJ2} then according to the voltage law the following equation holds:

$$v(t) = v_{\text{TJ1}}(t) + v_{\text{TJ2}}(t). \tag{1.1}$$

Now, we include the existence of an island charge (charge on the floating node):

$$q_{\text{i}}(t) = -q_1(t) + q_2(t). \tag{1.2}$$

Modeling the junction capacitance as a linear time-invariant capacitor and using its definition, $q_k(t) = C_{\text{TJ}k}v_{\text{TJ}k}(t)$ for each capacitor, $k = 1, 2$, we obtain:

$$\begin{cases} v(t) = v_{\text{TJ1}}(t) + v_{\text{TJ2}}(t) \\ q_{\text{i}}(t) = -C_{\text{TJ1}}v_{\text{TJ1}}(t) + C_{\text{TJ2}}v_{\text{TJ2}}(t). \end{cases}$$

Solving these two equations with two unknowns, we obtain:

$$v_{\text{TJ1}}(t) = \frac{C_{\text{TJ2}}}{C_{\text{TJ1}} + C_{\text{TJ2}}} v(t) - \frac{1}{C_{\text{TJ1}} + C_{\text{TJ2}}} q_{\text{i}}(t), \text{ and} \tag{1.3}$$

$$v_{\text{TJ2}}(t) = \frac{C_{\text{TJ1}}}{C_{\text{TJ1}} + C_{\text{TJ2}}} v(t) + \frac{1}{C_{\text{TJ1}} + C_{\text{TJ2}}} q_{\text{i}}(t). \tag{1.4}$$

We also obtain, using the capacitor definition:

$$q_1(t) = \frac{C_{\text{TJ1}}C_{\text{TJ2}}}{C_{\text{TJ1}} + C_{\text{TJ2}}} v(t) - \frac{C_{\text{TJ1}}}{C_{\text{TJ1}} + C_{\text{TJ2}}} q_{\text{i}}(t), \text{ and} \tag{1.5}$$

$$q_2(t) = \frac{C_{\text{TJ1}}C_{\text{TJ2}}}{C_{\text{TJ1}} + C_{\text{TJ2}}} v(t) + \frac{C_{\text{TJ2}}}{C_{\text{TJ1}} + C_{\text{TJ2}}} q_{\text{i}}(t). \tag{1.6}$$

Now, we consider the vi-relationship for each capacitor:

$$i_{\text{TJ}k}(t) = C_{\text{TJ}k} \frac{\mathrm{d}v_{\text{TJ}k}(t)}{\mathrm{d}t}. \tag{1.7}$$

For the current $i(t)$ through both capacitors C_{TJ1} and C_{TJ2} we obtain using equations 1.3 and 1.4:

$$i_{\text{TJ1}}(t) = \frac{C_{\text{TJ1}}C_{\text{TJ2}}}{C_{\text{TJ1}} + C_{\text{TJ2}}} \frac{dv(t)}{dt} - \frac{C_{\text{TJ1}}}{C_{\text{TJ1}} + C_{\text{TJ2}}} \frac{dq_i(t)}{dt}, \text{ and} \tag{1.8}$$

$$i_{\text{TJ2}}(t) = \frac{C_{\text{TJ1}}C_{\text{TJ2}}}{C_{\text{TJ1}} + C_{\text{TJ2}}} \frac{dv(t)}{dt} + \frac{C_{\text{TJ2}}}{C_{\text{TJ1}} + C_{\text{TJ2}}} \frac{dq_i(t)}{dt}. \tag{1.9}$$

If a current through this subcircuit is considered Kirchhoff's current law demands: $i_{\text{TJ1}}(t) = i_{\text{TJ2}}(t)$! Hence, the island charge $q_i(t)$ cannot be a continuous function of time; $q_i(t)$ *must be piecewise constant.* That is, only if we consider the island charge constant just before and constant just after tunneling we might be able to model the *non-tunneling* junction as a (charged) capacitor in a circuit theory.

What does the restriction means for a circuit description of tunneling? Considering tunneling to or from the island, we see that the island charge must change its value in zero time. Combining this with the knowledge that the voltage across the tunneling junction will step, because the voltage across the junction is a function of the island charge, we may guess that the tunnel current can be described by a (Dirac) delta pulse. The appearance of these delta pulses in capacitor networks will be investigated in the next section.

Concluding, the voltage across a junction in a structure consisting of two junctions and including a non-zero island charge is a function of both the voltage over the junctions and the island charge:

$$v_{\text{TJ1}}(t) = \frac{C_{\text{TJ2}}}{C_{\text{TJ1}} + C_{\text{TJ2}}} v(t) - \frac{1}{C_{\text{TJ1}} + C_{\text{TJ2}}} q_i(t), \text{ and}$$

$$v_{\text{TJ2}}(t) = \frac{C_{\text{TJ1}}}{C_{\text{TJ1}} + C_{\text{TJ2}}} v(t) + \frac{1}{C_{\text{TJ1}} + C_{\text{TJ2}}} q_i(t).$$

For non-zero island charges, these equations predict *nonlinear voltage-transfer characteristics*; in fact the characteristic is affine linear for non-zero island charges. This affine linear behavior, see figure Fig. 1.11, will be a topic in section 6.1.

1.4 Energy in Simple Capacitor Circuits, Bounded and Unbounded Currents

In the quantum physics description of the nanoelectronic devices energy plays an important role. This is a consequence of the appearance of energy in the Schrödinger equation (section 4.2). Because of this, it is expected

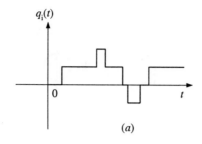

(a)

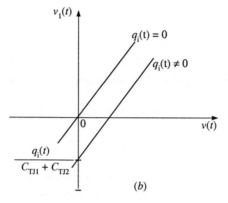

(b)

Fig. 1.11 (a): An example of the island charge $q_i(t)$; (b): Linear (only the line through the origin) and nonlinear voltage division in the two junction circuit in Coulomb blockade for values of the island charge $q_i(t) \neq 0$.

that energy will also be a key concept in the circuit description of the nanoelectronic devices.

In circuit theory, a circuit is characterized by one or more sources interconnected with one or more receivers, or sinks of electrical energy. In fact, in the circuit energy is supplied by sources, transferred from one place in the circuit to another, and temporarily stored or dissipated in circuit elements. Energy conservation in electrical circuits is treated in chapters 3 and 6. As a first step, energy in some simple circuits is considered.

1.4.1 *Switching circuits: energy in a resistive circuit*

Let us first discuss the energy in resistive dc circuits. Consider a circuit consisting of a passive linear resistor excited by an ideal dc-voltage source, see figure Fig. 1.12(1).

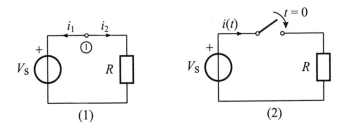

Fig. 1.12 In an energy context, the concept of a dc circuit is rather idealized. For example, here the energy dissipated in the passive linear resistor is infinite unless a switch in included.

In (electronic) circuit theory the circuit is described by the Kirchhoff current law (KCL), the Kirchhoff voltage law (KVL), and the constitutive relation of the linear resistor. The equations are, of course:

$$\text{KCL at node 1: } i_1 + i_2 = 0$$
$$\text{constitutive relation: } \quad i_2 \quad = R^{-1}v_R$$
$$\text{KVL : } \quad V_s \quad = v_R.$$

And they predict the typical behavior of resistive circuits.

In an energy context, the concept of a dc-voltage, however, is rather idealized. To illustrate this we calculate the energy absorbed by the linear resistor, w_R, and the energy delivered by the source $(-w_s)$:

$$w_R = R \int_{-\infty}^{t} i_2^2 \mathrm{d}\tau = \frac{V_s^2}{R} \int_{-\infty}^{t} \mathrm{d}\tau = \infty \qquad (1.10)$$

and

$$(-w_s) = -V_s \int_{-\infty}^{t} i_1 \mathrm{d}\tau = \frac{V_s^2}{R} \int_{-\infty}^{t} \mathrm{d}\tau = \infty. \qquad (1.11)$$

We see that energy is conserved, that is, the source provides the same amount of energy as is absorbed by the resistor, but the amount of energy is infinite.

This, of course, is not a realistic circuit description, because any circuit like this one will be turned on at a certain time. From the point of view of energy it is therefore more useful to described this circuit with inclusion of a switch, as is shown in figure Fig. 1.12(2).

To describe circuits that include switches we assume that the switch opens (or closes) at $t = 0$, we will distinguish $t = 0^-$ just before opening (closing) the switch, and $t = 0^+$ immediately after opening (closing) the switch. Defined are:

$$f(0^+) \stackrel{\text{def}}{=} \lim_{t \downarrow 0} f(t) \tag{1.12}$$

and

$$f(0^-) \stackrel{\text{def}}{=} \lim_{t \uparrow 0} f(t) \tag{1.13}$$

If the switch is closed at $t = 0$, the current will include a discontinuity at $t = 0$ in this new circuit:

$$i(t) = \frac{V_{\text{s}}}{R} \varepsilon(t), \tag{1.14}$$

in which $\varepsilon(t)$ is the unit step function defined by the relation

$$\varepsilon(t) = \begin{cases} 1 & \text{for} \quad t > 0 \\ 0 & \text{for} \quad t < 0. \end{cases} \tag{1.15}$$

At $t = 0$ the unit step function is undefined but restricted between 0..1. From a mathematics point of view the, in this way, defined step function is a "distribution" or "generalized function".

After closing the switch, that is for $t > 0$, the energy absorbed by the resistor in the circuit becomes:

$$w_{\text{R}}(t) = R \int_{-\infty}^{t} i^2 d\tau = \frac{V_{\text{s}}^2}{R} \int_{-\infty}^{t} \varepsilon^2(\tau) d\tau = \left[\frac{V_{\text{s}}^2}{R} \int_{0+}^{t} d\tau \right] \varepsilon^2(t)$$

$$= \frac{V_{\text{s}}^2}{R} t\varepsilon(t), \tag{1.16}$$

where is used that $\varepsilon^2(t) = \varepsilon(t)$, the fact that if the integrand is zero for $t < 0$, and that a resistor cannot dissipate in zero time (that is, during the switching action). Consequently, we can move the unit step function outside the integral, thereby changing its argument to t.

For the energy $(-w_{\text{s}}(t))$ delivered by the source we find:

$$(-w_{\text{s}}(t)) = -V_{\text{s}} \int_{-\infty}^{t} (-i) d\tau = \frac{V_{\text{s}}^2}{R} \int_{-\infty}^{t} \varepsilon(\tau) d\tau = \left[\frac{V_{\text{s}}^2}{R} \int_{0+}^{t} d\tau \right] \varepsilon(t)$$

$$= \frac{V_{\text{s}}^2}{R} t\varepsilon(t). \tag{1.17}$$

To evaluate what energy is delivered by the source at $t = 0$ we use the property of the source that in case of finite currents an energy source cannot deliver any energy in zero time; a property that is discussed in one of the coming subsections.

We, again, see that the energy is conserved in the circuit. In a similar way, the reader can check easily that for a resistor excited by an ideal current source conservation of energy holds too.

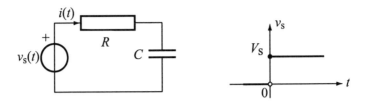

Fig. 1.13   Circuit for charging a capacitor

1.4.2   *Charging a capacitor: bounded current*

A reasonable circuit to model the charging of a linear capacitor is shown in Figure 1.13, if we assume that the resistance of the wire is substantial enough and the current is not changing very fast and always well defined— that is the current is always bounded. Supposing that the capacitor is discharged totally during $t < 0$, and it will start charging at $t = 0$. The capacitor will be charged with an amount q until the voltage across the capacitor equals V_s (steady-state). Kirchhoff's voltage law demands $v_s(t) = i(t)R + q(t)/C$.

Starting from energy conservation in the circuit, in the interval dt the voltage source delivers the energy dw and the circuit dissipates energy in the resistor and stores energy in the capacitor:

$$dw = V_s i dt = V_s dq = i^2 R dt + \frac{q}{C} dq. \tag{1.18}$$

The first term is the energy dissipation in the resistor, the second the amount of energy stored in the capacitor. In steady-state the voltage across the capacitor is $V_s = q_0/C$ and the current is zero. The total amount of energy absorbed by the circuit is:

$$\int_0^\infty i^2 R dt + \frac{1}{C} \int_0^{q_0} q dq =$$

$$w_R + \frac{1}{2C} q_0^2 = w_R + \frac{1}{2} q_0 V_s. \tag{1.19}$$

During charging, the source provided an amount of energy of $w_s = V_s \int_0^\infty i dt = q_0 V_s$, while the capacitor stored an amount of energy equal to $w_C = (1/2)CV_s^2 = (1/2)q_0 V_s$; we see that precisely half of the energy is dissipated in the resistor. This result holds for *any* non-zero value of the resistor, even for very small values. Notice that the source supplies additional energy to the energy to charge the capacitor and that energy is conserved in the circuit, because this additional energy is dissipated in the resistance.

1.4.3 *Charging a capacitor: unbounded current*

Equation 1.18 seems to indicate that in case of $R = 0$ the resistor term disappears and the source is providing twice as much energy as is stored on the capacitor, giving rise to the speculation that the other half of the energy is released as radiation in the circuit. To investigate this claim, consider the current in the circuit.

Staring with Kirchhoff's voltage law, $V_s = Ri + v_C$, and making use of the definition of the current $i(t) = dq(t)/dt$ we obtain

$$V_s - \frac{q}{C} = R\frac{dq}{dt} \quad \text{or} \quad V_sC - q = RC\frac{dq}{dt} \tag{1.20}$$

defining

$$(-q') = V_sC - q \tag{1.21}$$

then

$$q' = -RC\frac{dq'}{dt} \quad \text{or} \quad \frac{dq'}{q'} = -\frac{dt}{RC}. \tag{1.22}$$

The solution of this differential equation is

$$\int \frac{dq'}{q'} = \ln q' = -\frac{t}{RC} + \text{constant}$$

going back to q we obtain

$$q = \text{const.}e^{-t/RC} + V_sC. \tag{1.23}$$

For $t = 0$ the charge is $q = 0$, and we can determine the constant. We obtain

$$q = V_sC(1 - e^{-t/RC}). \tag{1.24}$$

Using $v = q/C$ we find the voltage across the capacitor

$$v_C = V_s(1 - e^{-t/RC}), \tag{1.25}$$

and using $i = dq/dt$ the current through the circuit

$$i(t) = \frac{V_s}{R}e^{-t/RC}. \tag{1.26}$$

Taking the limit $R \to 0$ the current in the circuit will become infinite. As the total charge stored on the capacitor is independent of R, the current can be expressed with the (Dirac) delta pulse and is called an unbounded

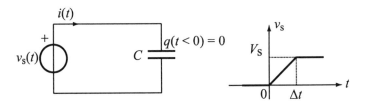

Fig. 1.14   Network to describe the charging of a capacitor with a finite current (bounded current) without a resistor.

current: $i(t) = CV_s\delta(t)$. (The definition used for the Dirac delta pulse, here, is

$$\delta(t) = \lim_{a \to \infty} ae^{-at}; \qquad (1.27)$$

other definitions also exist.)

An infinite current, however, is incompatible with the concept of resistance, in which charges must have a finite speed and time to move. In addition, the energy dissipation $w_R(t) = i^2(t)R$ in a resistor cannot be defined for infinite currents, because the square of a delta function is not defined in the theory of generalized functions.

To clarify energy calculations in capacitor networks without resistors we consider the network of figure Fig. 1.14. Note that this network describes really the charging of a capacitor without a resistor (the description of a circuit with even a very small resistor would fall within the previous description); circuits that are less realistic for modeling reality will be called networks from now on.

To analyze energy in the network of figure Fig. 1.14, we determine $v(t)$, $i(t)$, the energy stored on the capacitor w_{se}, and the total energy delivered by the (voltage) source w_s. The capacitor is defined by the relation $v = C^{-1}q$. Knowing the voltage as a function of time

$$v(t) = \begin{cases} 0 & \text{for} \quad t < 0 \\ (V_s/\Delta t)t & \text{for} \quad 0 \le t \le \Delta t \\ V_s & \text{for} \quad t > \Delta t \end{cases}$$

we can easily find the current in the circuit using

$$i(t) = C\frac{dv(t)}{dt} \qquad (1.28)$$

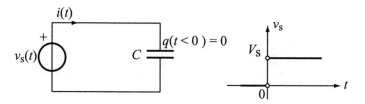

Fig. 1.15 Network to describe the charging a capacitor with an infinite current (unbounded current) without a resistor.

:

$$i(t) = \begin{cases} 0 & \text{for} \quad t < 0 \\ CV_{\text{s}}/\Delta t & \text{for} \quad 0 \le t \le \Delta t \\ 0 & \text{for} \quad t > \Delta t. \end{cases}$$

We can calculate the energy stored on the capacitor after $t > \Delta t$ to be $w_{\text{se}} = 1/2(CV_{\text{s}}^2)$. The energy delivered by the source in the interval $0 < t < \Delta t$ can be found as:

$$w_{\text{s}} = \int_0^{\Delta t} iv\,dt = \frac{CV_{\text{s}}^2}{\Delta t^2} \int_0^{\Delta t} t\,dt =$$

$$= \frac{1}{2} \frac{CV_{\text{s}}^2}{\Delta t^2} \Delta t^2 = \frac{1}{2} CV_{\text{s}}^2.$$

We notice that energy is conserved in this network.

To go back to our original challenge, the energy description in the capacitor circuit of Figure 1.13 in the limit $R = 0$, consider the limit $\Delta t \downarrow 0$, see figures Fig. 1.15 and Fig. 1.16. We can easily perform the calculations and find that also in this case the amount of energy delivered by the source is $w_{\text{s}} = 1/2(CV_{\text{s}}^2)$. That is, energy is conserved and there is no need for the introduction of energy loss by radiation. Also in this case the current can be described by a (Dirac) delta pulse, as is seen in figure Fig. 1.16. While Δt approaches zero, the current grows to infinity but the area in the ti-diagram stays the same, namely CV_{s}.

1.4.4 *Energy calculation with the generalized delta-function*

The mathematical definition of the Dirac delta function or impulse function $\delta(t)$ is:

$$\delta(t) \stackrel{\text{def}}{=} \begin{cases} 0 & \text{for} \quad t > 0 \\ \int_{-\infty}^{\infty} \delta(\tau)d\tau = 1 & \text{for} \quad t = 0 \\ 0 & \text{for} \quad t < 0. \end{cases} \tag{1.29}$$

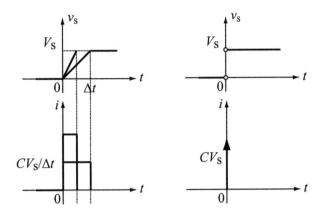

Fig. 1.16   Voltages and currents in circuits of figures Figs. 1.14 and 1.15.

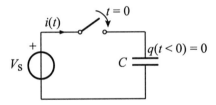

Fig. 1.17   Switched capacitor network with an unbounded current when the switch is closed.

The $\varepsilon(t)$ and $\delta(t)$ are not regular analytic functions but generalized functions. They define each other:

$$\int_{-\infty}^{t} \delta(\tau)\mathrm{d}\tau = \varepsilon(t) \tag{1.30}$$

$$\frac{\mathrm{d}}{\mathrm{d}t}\varepsilon(t) = \delta(t) \tag{1.31}$$

Using the expressions for the step and delta function one finds for the current $i(t)$, $i(t) = C\mathrm{d}v/\mathrm{d}t = CV_s\mathrm{d}(\epsilon(t))/\mathrm{d}t = CV_s\delta(t)$. The delta current pulse $CV_s\delta(t)$ that transports in zero time an amount of charge $Q = CV_s$ is, generally, called an **unbounded current**.

We can also calculate directly the energy provided by the source by noting that while the source steps from 0 V to V_s the current through the source is the delta pulse $CV_s\delta(t)$. The energy provided by the source w_s is

$$|w_s| = \int_{-\infty}^{\infty} iv\mathrm{d}t = CV_s^2 \int_{-\infty}^{\infty} \delta(t)\epsilon(t)\mathrm{d}t = \frac{1}{2}CV_s^2. \tag{1.32}$$

Where we used that the integral can be calculated by partial integration and realizing that the delta pulse is defined as the derivative of the step function:

$$\int_{-\infty}^{\infty} \delta(t)\epsilon(t)\mathrm{d}t = \frac{1}{2}[\epsilon^2]_{-\infty}^{\infty} = \frac{1}{2}. \tag{1.33}$$

The important difference with the circuit described in figure Fig. 1.13 is the absence of the resistor.

It is important to note that, based on energy arguments, the previous circuit shows that an impulsive current is not compatible with the presence of a resistor in the circuit. If a resistor would have been included the source had had to provide twice as much energy; which is not possible with a delta pulse current. Finally, we notice that the network of figure Fig. 1.15 represents a switched network, as shown in figure Fig. 1.17.

1.5 Operational Temperature

Single-electron electronics will become possible if circuits are considered that include tunneling capacitors in Coulomb blockade. In this way it is possible to store and retrieve additional electrons on islands. For tunneling, the separation of the island and the metallic lead towards it must be very small. In spite of this, the capacitance of the junction must be very small too. The reason for this is the influence of temperature. Although the role of temperature will be discussed later on, the effects can be estimated roughly by considering energy scales.

Until now, the effect of non-zero temperature is not taken into account. For example in figure Fig. 1.8 it is assumed that there are no electrons above the Fermi levels nor empty levels below the Fermi levels. This is only possible in the limit $T = 0$ K. At temperature T the probability $f = f(E,T)$ that a state at energy E is occupied by an electron is given by the **Fermi-Dirac distribution**

$$f = \left[\exp(\frac{E - E_\mathrm{F}}{kT}) + 1\right]^{-1}, \tag{1.34}$$

k the Boltzmann constant, and the probability $g = g(E,T)$ that an energy state is empty is

$$g = 1 - f. \tag{1.35}$$

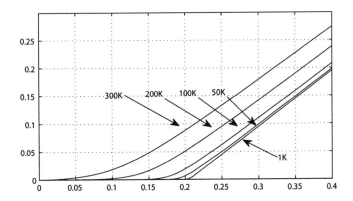

Fig. 1.18   Example of calculated behavior of a single tunnel junction: different curves for various temperatures according to the impulse circuit model. On the horizontal axis the voltage across the junction, on the vertical axis the tunnel current (normalized) through the junction; the critical voltage in his example is 0.2V and the tunneling time is taken zero.

In the impulse circuit model the critical voltage is found by determining the voltage across the tunnel junction for which holds that the voltage after the tunnel event equals minus the voltage before the tunnel event, as such imposing energy conservation. This evaluation is done for electrons at the Fermi level; the physics condition belonging to this is a description at $T = 0$ K. Operation at nonzero temperatures will therefore result in possible tunnel events before the critical voltage is reached, as can be seen in figure Fig. 1.18. To use or to observe Coulomb blockade at higher temperatures (say, upto room temperatures) the critical voltage must be large enough.

We can use this knowledge to get an idea of the values of junction capacitances that may result in an observable Coulomb blockade at moderate temperatures. The maximum value of the critical voltage of a tunnel junction is (chapter 7):

$$v_{\mathrm{TJ}}^{\mathrm{cr}}(\max) = \frac{e}{2C_{\mathrm{TJ}}}.$$

As an example assume that $v_{\mathrm{TJ}}^{\mathrm{cr}}(\max) = 0.2$ V we find that the junction capacitance must be smaller than 40 aF. *Obviously we have to use nanometer structures.* Figure 1.19 shows a realistic small metal-insulator-metal-insulator-metal structure.

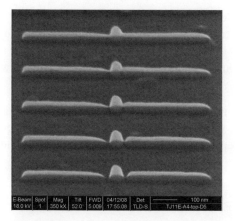

Fig. 1.19   State of the art (2008) example of two junctions and an island. The gap in the lowest structure is about 10 nm wide and the lines have width and hights of about 20 nm. SEM picture; the structures were fabricated using electron beam induced deposition (EBID); the metal lines and the island are made of chromium. [Heerkens *et al.* (2008)]

Example 1.1 Let's assume that we have two metallic leads like the bars in figure Fig. 1.1. Suppose such a arrangement could be manufactured with $d = 7$ nm, $h = 4$ nm, and $w = 10$ nm. The dielectric constant of the insulating material κ is 9. Can we get an idea of the capacitance of this structure?

$\triangledown$ A is the area of each lead and $\epsilon_0 = 8.85 \times 10^{12}$ F/m is the permittivity of free space. If we use a simplified model neglecting edge effects, the capacitance of the two parallel plates is given by:

$$C = \frac{\kappa \epsilon_0 A}{d} = \frac{(9)(8.85 \times 10^{-12} \text{F/m})(4 \times 10^{-9} \text{m})(10 \times 10^{-9} \text{m})}{(7 \times 10^{-9} \text{m})} = 0.45 \text{aF}.$$

$\triangle$

1.6   Research Questions

This monograph provides an introductory text to modern nanoelectronic circuit design. As a text for senior undergraduate and graduate students it is closely related to modern nanoelectronic research. At the end of the introductory chapter it is, therefore, useful to formulate the research questions we try to answer. In general, those questions are:

(1) What concepts from solid-state or semiconductor physics should be

used for the description of nanoelectronics?

(2) Can nanoelectronics be described with models familiar in electrical circuit theory?

(3) What does nanoelectronics learn us about concepts we use in classical circuit theory?

As we are just entering the nanoelectronic era, these questions must be solved as soon as possible. This text is just a starting point. By deriving a circuit model for the tunneling junction it makes a modest step into the future.

Problems and Exercises

1.1 The device properties of a nanoelectronic device are described by a process that minimizes energy. On what level(s) can this device be modeled?

1.2 Various equivalent circuits have been proposed for metallic SET junctions. Which of them model the tunneling capacitor as a leaky capacitor?

1.3 Suppose that after tunneling an electron becomes a hot electron. If you want to model the energy loss of a hot electron by a circuit element, what circuit element would you choose, and why?

1.4 Consider the equivalent subcircuit for two tunnel junctions in series in Coulomb blockade. Show that if the island charge is q_i charge neutrality in the total circuit is still maintained.

1.5 Show that energy is conserved in a circuit in which a resistor is excited by an ideal current source.

1.6 Show that the voltage across an initially uncharged capacitor excited by a delta-pulse current $i = e\delta(t)$ steps from 0 to e/C Volt.

Chapter 2

Tunneling Experiments in Nanoelectronics

Nanoelectronics can be seen as an experimental science. Based on quantum physical phenomena, researchers design nanoelectronic components. Subsequently, they make the devices, measure them, and interpret the measurements. Knowing the electrical device characteristics we, then, try to analyze and synthesyze nanoelectronic circuits and ditto systems.

In general, specific experiments and their interpretations are not covered in text books. To large extend, this situation is unavoidable, and even proper. It would be impractical to describe in significant detail all the experiments that lead to a certain development, like nanoelectronics. To bring some order in the sometimes confusing area of new nanoelectronic devices, however, it seems appropriate to present and discuss some landmark experiments, investigations that mark a turningpoint or an important stage in our ideas.

The choice of experiments, included in this text, is to some extend arbitrary. Selected are those that form a chain from the first experiment showing the quantum mechanical tunneling phenomenon to experiments showing Coulomb blockade, that is, no tunneling is observed while tunneling events were expected to happen.

2.1 Tunneling in the Tunnel Diode

First the experiments of Esaki on the tunnel diode and their interpretations are presented. The experimental findings are explained by introducing quantum mechanical tunneling. As an introduction to the tunnel diode, first, the energy-band diagram of the ordinary diode is discussed with a focus on the role of the electrons. Subsequently, the tunnel diode and experiments on the tunnel diode are treated. To explain the characteristic in the vi-plane quantum mechanical tunneling is proposed: electrons tunnel to empty energy levels. After this, an equation for the tunnel current is derived when the applied voltage is small. Returning to electronics, a circuit interpretation of the nonlinear voltage-current characteristic of the tunnel diode is discussed. The discussion will lead to the necessity for a precise definition of energy dissipation before, during, and after the tunneling event, if present, in an electronic circuit. The section ends with a short discussion on the resonant tunneling device, a quantum electronic device more recently developed. Before we start with describing the tunnel diode it is worth mentioning that this device is recently gaining attention once more due to new possibilities for the integration [Wernersson *et al.* (2005)] into very large scale integration (VLSI) circuits.

2.1.1   *Energy-band diagram for the p − n diode*

The contact (junction) between an n-type and p-type semiconductor can be described qualitatively. The **n-type** crystal contains a large number of electrons and the **p-type** a large number of holes. Figure 2.1 shows that after the contact has been made, a depletion region will be formed. Electrons diffuse from the n-type part into the p-type part, and holes diffuse from the p-type part into the n-type part. The mobile carriers recombine leaving behind a positive space charge in the n-type part and a negative space charge in the p-type part, in this way building up an electric field opposing the diffusion of electrons from the n-side and holes from the p-side. This electric field will cause a drift current of minority carriers near the junction. Electrons in the the p-side will drift to the n-side, and holes in the the n-side will drift to the p-side. Equilibrium is reached when the opposing field becomes sufficiently large to reduce both net electron and net hole currents to zero, with the diffusion and drift components equal in magnitude but oppositely directed. The region near the junction where the mobile carrier concentrations have been reduced below their thermal equilibrium values is called the **depletion region**.

If $\mathcal{E}$ is the electric field and V the potential then the actual potential distribution around the junction can be calculated with the aid of Poisson's equation:

$$\frac{\mathrm{d}\mathcal{E}}{\mathrm{d}x} = -\frac{\mathrm{d}^2V}{\mathrm{d}x^2} = \frac{\text{space-charge density}}{\epsilon\epsilon_0}. \qquad (2.1)$$

Semiconductor junctions, now, have two major properties: (1) there is a

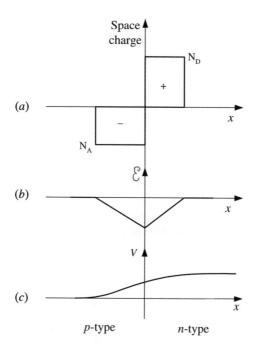

Fig. 2.1   Schematic plots of space charge (a), electric field (b), and potential variations (c) around the *p-n* junction.

space charge and an electric field across the junction; thereby the junction is **pre-biased** (i.e., there is a built-in potential); the pre-biased condition can be maintained indefinitely. (2) The impurity atoms maintaining the space charge are immobile in the temperature range of interest.

If we apply an external voltage that increases the voltage across the junction, it also increases the electric field; we do not expect that under these conditions a significant current will flow. If we, however, reduce the voltage, there will be a large current as holes and free electrons flow into opposite regions. In the former case, the junction is said to be **reverse-biased**, and **forward-biased** in the latter case. This qualitative discussion already implies that this *p-n* junction is a diode. It conducts very well in one direction (forward bias), but very poorly in the other direction (reverse bias). Because of recombination and generation of carriers, not discussed here, there will be a small current if the diode is reverse-biased.

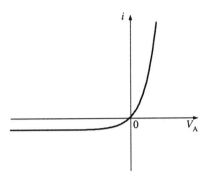

Fig. 2.2   Voltage-current characteristic *p-n* diode

Based on the qualitative description the so-called ideal diode equation can be derived. The result is well-known:

$$i = I_0 \left\{ \exp\left(\frac{eV_A}{kT}\right) - 1 \right\}, \qquad (2.2)$$

in which e is the fundamental charge, k the Boltzmann constant, and T the temperature. The characteristic shown in figure Fig. 2.2.

The depletion region represents a parallel-plate capacitor too, because for a *small* voltage increase we add charges at the boundaries. This is the so-called **junction capacitance** C_J and depends on the applied voltage V_A, dielectric constant, area, and doping levels:

$$\frac{dq}{dV_A} = C_J. \qquad (2.3)$$

So far we have considered only the spatial distribution of carriers within the diode. We did not consider the distribution of carriers in energy. Such a discussion leads to energy-band diagrams, and profide us clues for the differences of semiconductors, metals, and insulators. It also explains the difference between the "normal" diode and the tunnel diode.

To apply an energy band model the Fermi-energy level E_F must be a constant and independent of the position for any junction at thermal equilibrium. For an *n*-type semiconductor, the Fermi level is above the middle of the bandgap; for *p*-type, it is below. If we contact the *p*- and *n*-type semiconductors the Fermi levels always align in thermal equilibrium when we do not apply an external voltage (bias). For the electron distribution is controlled by the *same* Fermi-Dirac statistics. In thermal equilibrium, where no net electron current can flow, the probability must be the same

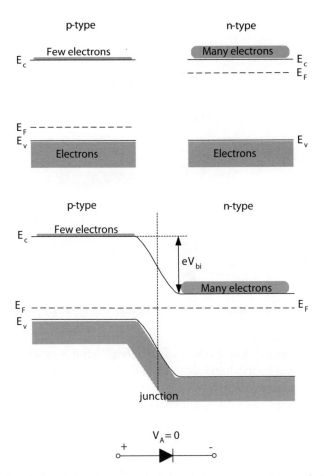

Fig. 2.3 Alignment of the Fermi levels when *p*- and *n*-type regions are contacted in thermal equilibrium (zero bias). The barrier is formed due to the band displacement. Only the areas with electrons are shown; not shown are the many holes in the valence band of the *p*-type and the few holes in the valence band of the *n*-type semiconductor.

at any given energy level throughout the semiconductors. (If they were not there would be a current flowing!) This argument also holds for the Fermi levels.

The energy band under equilibrium conditions is illustrated in figure Fig. 2.3. Since E_F is a constant at thermal equilibrium, the band edges (E_c and E_v) have to change their position relative to E_F going from *p*- to *n*-material. The slope of the energy band edges is proportional to the electric field; in this case a negative slope yields a negative electric field in

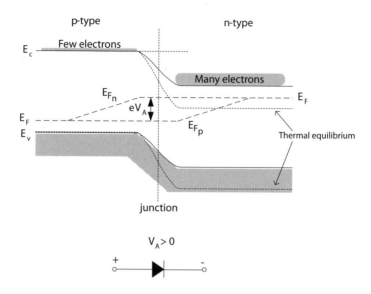

Fig. 2.4   Energy diagram of an *p-n* junction with applied forward bias

the depletion layer. The curvature of the energy band is proportional to the charge density. The left-hand side of the depletion region with its negative curvature implies a negative charge density in the p-material, while the positive curvature indicates a positive charge density in the n-material.

The effect of a reverse bias is simply that the slope of the band edges in the depletion region has increased, reflecting the increase in electric field. Also the barrier height is increased. The diffusion currents are reduced, while the drift currents remains at their thermal equilibrium values. A small reverse current results.

Applying a negative potential to the n-type region with respect to the p-type region, that is, applying a forward bias ($V_A > 0$) is illustrated in figure Fig. 2.4. The barrier is lowered and majority carriers pour into the opposite regions. A large number of electrons, from the n-type region will diffuse into the p-region causing a positive current, and are called **injected electrons** once they reach the p-region bulk. The number of electrons that drift down the potential hill remains equal to its (small) thermal equilibrium value. Arguments for holes are similar to those for electrons. The net effect of forward bias is a large increase in the diffusion current while the drift components again remain fixed near their thermal equilibrium values.

The majority carriers will appear as minority carriers after injection which increases the minority carrier concentration exponentially with the applied voltage at the depletion-layer boundaries. An interesting aspect of the forward-biased junction is the splitting of the Fermi-level for electrons and holes, due to the nonequilibrium conditions. As a matter of fact, in order to maintain space-charge neutrality outside the depletion layer, equal quantities of excess majority carriers are drawn in to compensate the excess minority carriers; both excess carriers do not move independently of each other, a phenomenon called ambipolar transport. The process results in a relatively insignificant alteration in the majority carrier concentrations, but a large change in the minority carrier concentrations. In order to describe these alterations in terms of the Fermi distributions, we must use different Fermi levels for the electrons and holes. These levels are called *quasi-Fermi levels* because they are used in nonequilibrium conditions.

So, due to the nonequilibrium conditions of the junction in forward bias (a current is flowing!) the Fermi-levels split. In the bulk there is only one Fermi level, because thermal equilibrium prevails. Approaching the junction from the p-side, a large increase in the electron concentration takes place. Consequently, the electron quasi-Fermi level rises, while the hole quasi-Fermi level remains essentially unchanged. As for the depletion region, the quasi-Fermi levels remain constant, since we are using the assumption that the concentrations at the two boundaries are related exponentially with the voltage, and the bands follow the potential distribution. In the n-region, a convergence of the quasi-Fermi levels can be seen as the excess hole concentration disappears. (Note that the exponential decay of excess carriers is represented by a practically linear variation in the quasi-Fermi levels.) Electrons from the outside circuitry recombine with the injected holes in the n-type region causing the Fermi-level to change. Outside this region of recombination the Fermi-level is constant again (thermal equilibrium); however, the Fermi-levels at both sides are separated by eV_A.

In essence, since the Fermi-Dirac function distributes the carriers nearly exponentially with increasing energy, one expects the number of carriers able to diffuse to increase exponentially with the reduction of the potential barriers. This being the case, the net forward bias current should increase exponentially with V_A.

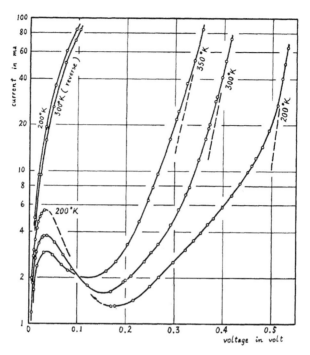

FIG. 1. Semilog plots of the measured current-voltage
characteristic at 200°K, 300°K, and 350°K.

Fig. 2.5   Original figure of Leo Esaki [Esaki (1958)].

2.1.2   *Experiments of Esaki on the (tunnel) diode*

In 1957, Leo Esaki's small research group at the Sony Corporation in Tokyo
looked into diodes with degenerately doped germanium *p-n* junctions. In
the heavily doped germanium *p-n* junctions, Esaki discovered an "anomalous" nonlinear voltage-current characteristic when the junction was biased
in the forward direction, that is, a *negative differential-resistance* region was
found over part of the *vi*-characteristic when a *forward bias* was applied, besides this the diode appeared to be more conductive when a reverse bias was
applied than when a forward bias was applied. The obtained characteristic
was explained by the quantum tunneling concept and qualitative agreement
was obtained between the tunneling theory and the experiment results. The
group created what can be called the first nanoelectronic device. For this
experimental discovery of tunneling in semiconductors he shared the 1973
Nobel Prize for physics with Ivar Giaever and Brian Josephson.

Figure 2.5 shows the voltage-current characteristic obtained by Esaki. "In the course of studying [...] very narrow germanium *p-n* junctions, we have found an anomalous voltage-current characteristic in the forward direction [...]. In this *p-n* junction [...] the acceptor concentration in the *p*-type side and the donor concentration in the *n*-type side are, respectively, $1.6x10^{19}$cm^{-3} and approximately 10^{19}cm^{-3}. The maximum of the curve was observed at 0.035 ± 0.005 volt in every specimen. [...] In the range over 0.3 volt in the forward direction, the voltage-current curve could be fitted almost quantitatively by the well-known relation: $I = I_s[\exp(qV/kT) - 1]$. This junction diode is more conductive in the reverse direction than in the forward direction. [...] the junction width is approximately 150 angstrom (15nm) [...] Owing to the large density of electrons and holes, their distribution should become degenerate; the Fermi level in the *p*-type side will be 0.06 eV below the top of the valence band, and that in the *n*-type will lie above the bottom of the conduction band [...]. " [Esaki (1958)]

For small voltages, the *vi*-characteristics are reasonably well explained by assuming that electrons *tunnel* through a small barrier to empty states at the same energy level, causing the electron to follow a horizontal path in the energy-band diagram. It is because this tunneling effect that the heavily doped diode with a junction of only 10 nm wide is called **Esaki diode** or **tunnel diode**.

2.1.3 *Nonlinear voltage-current characteristic of the tunnel diode*

The tunnel diode is a semiconductor device that consists of a heavily doped abrupt *p-n* junction. The doping on the opposite sides of the junction is such that the concentration of doping impurities exceeds the density of states in the conduction band in the *n*-type semiconductor and the density of states in the valence band in the *p*-type semiconductor. As a consequence the Fermi level in the *n*-type semiconductor lies within the conduction band, and in the *p*-type semiconductor in the valence band.

The space-charge or depletion region is extremely narrow in this structure, and forms a natural barrier for diffusing electrons or holes. In contrast to the conventional diode, in which the barrier is reduced at forward bias and injection begins *above* the barrier, the operation of the tunnel diode is explained by tunneling of electrons *through* the barrier at small forward-bias voltages. We will now consider the behavior of the tunneling junction

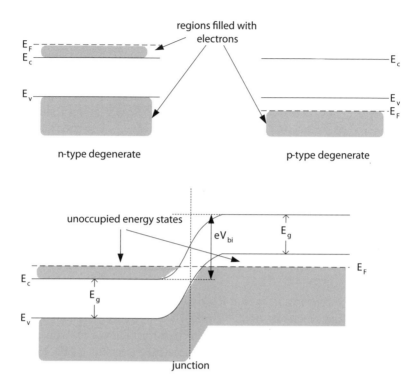

Fig. 2.6   Energy-band diagrams for degenerately doped *n*-type and *p*-type semiconductors, and the energy-band diagram at the interface of the tunnel diode

at low temperatures for convenience; at low temperatures all states are filled up with electrons to the Fermi level, while above it most states are empty. Figure 2.6 shows the energy-band diagram of the tunnel diode in thermal equilibrium. The operation of the tunnel diode can be explained, qualitatively, using diagrams at increasing forward-bias as in figure Fig. 2.7.

We start at zero bias voltage, see also figure Fig. 2.6. If we assume that we are near 0 K, then all energy states are filled with electrons below E_F on both sides of the junction. So, there are no empty states available to which the electrons can go. Therefore, *the tunnel current is zero when no bias voltage is applied*. Note that, at temperature T the probability of states being occupied by electrons is given by the Fermi-Dirac distribution function. So, for temperatures above absolute zero, there is a nonzero probability that some energy states above E_F will be occupied by electrons and some energy states below E_F will be empty. Tunneling will be possible.

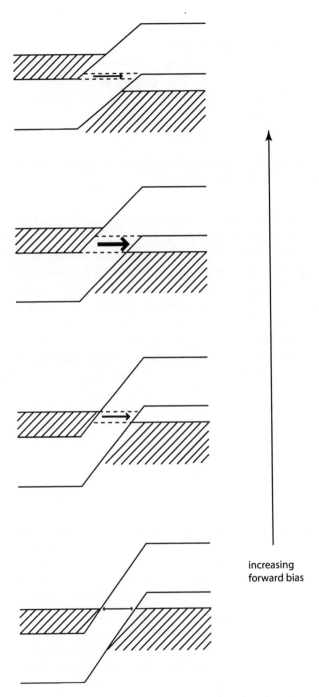

Fig. 2.7 Energy-band diagrams with electron tunneling flow in a tunnel diode with increasing forward-bias.

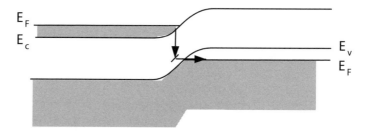

Fig. 2.8   Energy-band diagram schematically illustrating a possible dissipative tunneling event via a state in the forbidden gap.

However, now, the *net* tunnel current by electrons from the valence band to empty states in the conduction band and from the conduction band to empty states in the valence band should be zero, because no current flows through the circuit. Now, we apply an increasing forward-bias.

Figure 2.7 shows four diagrams in order to demonstrate what happens. If the junction is slightly forward-biased. Some of the electrons in the *n*-region face empty states in the *p*-region. *Note that according to our model during tunneling the total energy of the electron is unchanged, causing the electrons to follow a horizontal path in the energy-band diagram.* This will we call **direct tunneling**. Tunneling of electrons from the *n*-region into the *p*-region will follow, and there will be a small current. As the forward-bias voltage continues to increase, the number of electrons facing opposite empty states increases. A maximum tunnel current will be obtained when the maximum number of electrons face the maximum number of empty states. After the maximum tunnel current has been achieved the number of electrons facing empty states will be decreasing because there are no possible states in the forbidden gap, and the tunnel current will be become smaller.

With still increasing bias-voltage an increasing number of electrons will face the bandgap in the *p*-region. The tunnel current of electrons to empty states at the *same* energy level will approach zero, but the forward-bias will become large enough such that two other phenomena take over the behavior of the tunneling junction. The first is called **excess current**, and is mainly due to tunneling by way of energy states within the forbidden bandgap. In this type of tunneling electrons start in the conduction band, tunnel to a single energy state or in steps to multiple energy states in the forbidden bandgap, and finally tunnel to an empty state in the valence band. Tunneling in this way is now a dissipative process. Figure 2.8 shows the

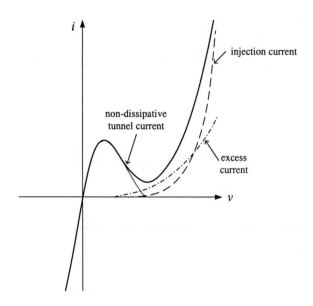

Fig. 2.9 vi-characteristic of a tunnel diode broken down into the three basic current components

basic mechanism. The second phenomenon is just the injection of electrons *over* the barrier becomes possible (in the same way as this happens in a normal diode).

The nonlinear voltage-current characteristic is now determined by the sum of the tunnel current, the excess current, the thermal injection current, and is plotted in figure Fig. 2.9. The large current in the reverse-bias region indicates an electron tunnel current too, but now from the the p-type side into the n-type side. In general, the probability for indirect tunneling is much lower than the probability for direct tunneling when direct tunneling is possible [1].

Example 2.1 Consider a silicon $p - n$ junction at $T = 300$ K with doping concentration of $N_d = N_a = 5 \times 10^{19} \text{cm}^{-3}$. Assuming the abrupt junction approximation is valid, determine the depletion layer width at a forward-bias voltage of

---

[1]A more complete description of tunneling includes (1) "direct tunneling, "phonon-assisted tunneling, and "indirect tunneling through inter-band states; see for example [Sze (1981)]. For our later discussion of tunneling between two metals it is enough to consider only elastic (nondissipative) direct tunneling.

$V_A = 0.40V$.

▽ The dielectric constant of silicon is 11.7, $\epsilon_0 = 8.85 \times 10^{12}$ F/m is the permittivity of free space, and the intrinsic carrier concentration of silicon is 1.5×10^{10} cm^{-3}. If we use the well-known semiconductor physics equations for the built-in voltage and the space charge width (depletion layer width) at temperature T:

$$V_{\text{bi}} = \frac{kT}{e} \ln\left(\frac{N_a N_d}{n_i^2}\right) \tag{2.4}$$

and

$$W = \left[\frac{2\epsilon(V_{\text{bi}} - V_A)}{e}\left(\frac{N_a + N_d}{N_a N_d}\right)\right]^{1/2} \tag{2.5}$$

we find $V_{\text{bi}} = 1.14$ V and thus $W = 6.2$ nm.

△

2.1.4   *Tunnel current*

When the barrier is sufficiently small, a finite probability exists for tunneling through the forbidden bandgap. The tunneling of an electron through the gap is formally the same as a particle tunneling through a potential barrier. As will be discussed in chapter 5, barriers with different shapes have different tunneling probabilities $\mathbf{T}(E)$.

The nondissipative tunneling current in the tunnel diode can be formulated in the following way; see also chapter 4. The number of states per unit energy range at any particular energy E occupied by electrons *in thermal equilibrium* at a given temperature T is (see section 4.5.4):

$$N(E) = 2f_{\text{F}}(E)\tilde{g}(E). \tag{2.6}$$

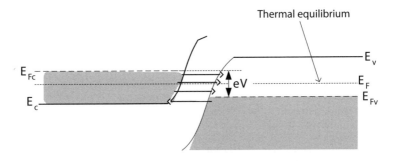

Fig. 2.10   Tunnel current at small forward bias

The factor of 2 takes into account the fact that there are two allowed directions for electron spin, $\tilde{g}(E)$ is the density of states per spin direction, and $f_F(E)$ is the Fermi-Dirac distribution. We designate $f_c(E)$ and $f_v(E)$ as the Fermi-Dirac distributions in the conduction band and valence band, respectively. At zero bias, the tunneling current $i_{c\rightarrow v}$ from the conduction band to an empty state (at the same energy level) in the valence band and the current $i_{v\rightarrow c}$ from the valence band to empty states in the conduction band are:

$$i_{c\rightarrow v} = A \int_{E_c}^{E_v} f_c(E)\tilde{g}_c(E)\mathbf{T}_{c\rightarrow v}(E)\{1 - f_v(E)\}\tilde{g}_v(E)\mathrm{d}E$$

$$i_{v\rightarrow c} = A \int_{E_c}^{E_v} f_v(E)\tilde{g}_v(E)\mathbf{T}_{v\rightarrow c}(E)\{1 - f_c(E)\}\tilde{g}_c(E)\mathrm{d}E. \qquad (2.7)$$

When the junction is slightly biased positively or negatively, the observed tunnel current can be found by combining both tunnel currents. Assuming that the tunneling probabilities can be considered equal, we find using:

$$f_c(E)\{1 - f_v(E)\} - f_v(E)\{1 - f_c(E)\} =$$

$$f_c(E) - f_c(E)f_v(E) - f_v(E) + f_v(E)f_c(E) =$$

$$f_c(E) - f_v(E),$$

for the tunnel current:

$$i = i_{c\rightarrow v} - i_{v\rightarrow c} = A \int_{E_c}^{E_v} \{f_c(E) - f_v(E)\}\mathbf{T}(E)\tilde{g}_c(E)\tilde{g}_v(E)\mathrm{d}E. \qquad (2.8)$$

If we, in addition to this, also assume the density of states on both sites equal and constant we can explain the linear behavior of the v-i characteristic at small values of the forward or reverse bias. With the assumptions, the tunnel current can now be written as:

$$i = A' \int_{E_c}^{E_v} \{f_c(E) - f_v(E)\}\mathbf{T}(E)\mathrm{d}E, \qquad (2.9)$$

It is clear that $i = 0$ when $v = 0$ and thus $E_{Fc} = E_{Fv}$.

For sufficiently small bias voltages and neglecting electrons tunneling from the p-side to the n-side, we note that all electrons at the n-side between Fermi level of the n-side (E_{Fc}) and the Fermi level of the p-side (E_{Fv}) can tunnel, because those electrons face empty states at the p-side. Consequently, the tunnel current will be proportional to the this difference. We know that the difference between these levels equals e times the applied voltage, thus, the current will be proportional to the applied bias voltage. That is, a linear resistive behavior exists; as is seen in the original experiment of Esaki, figure Fig. 2.5.

The importance of this conclusion is that the model, in which we only consider direct nondissipative tunneling, causing the electrons to follow a horizontal path in the energy-band diagram is a valid model for the tunneling in the tunnel diode at small bias voltages. A second important conclusion is that the *vi*-characteristic over the full voltage domain is nonlinear, this behavior paves the way to interesting electronic applications.

2.1.5   *Energy paradox*

As the *vi*-characteristic in Figure Fig. 2.9 shows, the tunnel diode is a nonlinear resistor. Simple interpretation of this figure also shows that the tunnel diode is a dissipating device; the instantaneous power entering the tunnel diode is always positive (although the differential resistance is sometimes negative). In a circuit the dissipated power is balanced by the power delivered by the source(s). In case of band-to-band tunnelling current and excess current we may consider *individual* electrons, because both currents involve the tunnelling of single electrons; at sufficiently low bias voltages the tunnel current can be described as composed of individual tunnelling events that are uncorrelated.

First consider the excess current. This excess current is mainly due to tunnelling by way of energy states within the forbidden bandgap. In this type of tunnelling, electrons start in the conduction band, tunnel to a single energy state or in steps to multiple energy states in the forbidden bandgap, and finally tunnel to an empty state in the valence band. Indirect tunnelling in this way is a dissipative process.

Now, secondly, consider the direct band-to-band tunnelling. For small voltages, the *vi*-characteristics are reasonably well explained by assuming that electrons tunnel through the small barrier to empty states at the same energy level, causing the electron to follow a horizontal path in the energy-band diagram. Actually, during this tunnelling energy is conserved; so there is no energy dissipation during the tunnel event. However, according to the nonlinear *vi*-characteristic the device is dissipating; where is the energy? To solve this paradox, energy has to be dissipated outside the tunnel region.

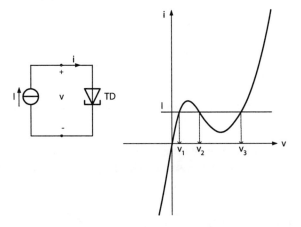

Fig. 2.11 More solutions for a tunnel diode excited by an ideal current source.

2.1.6 *Equivalent circuit*

As the *vi*-characteristic in figure Fig. 2.9 shows the tunnel diode is a resistor. It is clearly not an ohmic resistor. Due to the nonlinear *vi*-characteristic we can at best call it a nonlinear resistor. This nonlinear resistor is voltage-controlled; the current i is a (single-valued) function of the voltage v. By virtue of the nonlinearity of the tunnel diode a network including the tunnel diode can have more solutions.

An example is shown in figure Fig. 2.11. Although this circuit is a nonvalid model, the inclusion of a capacitor (or junction capacitance) will change the circuit to a dynamic one having two stable points (v_1, I) and (v_3, I); the point (v_2, I) is unstable. The resulting circuit can be switched between the two stable points.

The characteristic clearly shows the behavior of the tunnel diode as a nonlinear passive resistance. Under small-signal conditions, however, the tunnel diode can be biased into a negative resistance region around the *quiescent* or *operating point* Q, in which the *small signal* or *differential resistance*, $R_Q = dv/di$, is negative. This makes the tunnel diode an active (linear) element in this region and power gain is available for small AC signals.

Now, the electronic behavior of the tunnel junction can be examined by considering its equivalent circuit. The usual equivalent circuit is shown in figure Fig. 2.12, it consists of four elements: the series inductance L_s, the series resistance R_s, the diode capacitance C, and the negative diode

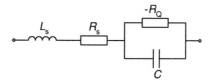

Fig. 2.12 High-frequency equivalent circuit for a tunnel diode biased into the negative resistance region.

resistance $-R_Q$. Note the parallel combination of the diode capacitance and the tunnel resistance (biased in the negative resistance bias). The impedance of the equivalent circuit of figure Fig. 2.12 is:

$$Z = \left[R_s - \frac{R_Q}{1 + (\omega R_Q C)^2} \right] + j \left[\omega L_s - \frac{\omega R_Q^2 C}{1 + (\omega R_Q C)^2} \right]. \qquad (2.10)$$

From equation 2.10 we can see that the resistive (real) part of the impedance will be zero at a certain frequency, and the reactive part will be zero at a second frequency. Generally these frequencies are denoted as the resistive cut-off frequency f_r, and the reactive cut-off frequency f_r, respectively. For the resistive cut-off frequency we obtain:

$$f_r \stackrel{\text{def}}{=} \frac{1}{2\pi R_Q C} \sqrt{\frac{R_Q}{R_s} - 1} \qquad (2.11)$$

If the diode is biased in the negative resistance region, R_Q is typically $-20\ \Omega$ and the resistive cut-off frequency is very high. Since the negative resistance does not decrease except at very high frequencies, the frequency response is determined by the $-R_Q C$ time constant. This time constant may be as low as 10^{-10} s or lower, allowing amplification (or oscillation) at high frequencies. When used in digital applications, the switching speed of the device is also in the order of this time constant, i.e., less than 1 ns.

2.1.7 *Resonant tunneling diode*

The heavily doped $n^{(+)} - p^{(+)}$ junction was an obstacle for integration of the tunnel diode into modern silicon IC technology for long time. Therefore, from 1980 systematic research was started on a modern device that can be seen as a follow up of the tunnel diode: the nanoelectronic negative resistance device called resonant tunneling diode (RTD). The basic device configuration is a five layer double quantum barrier structure, as shown in figure Fig. 2.13.

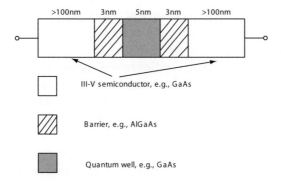

Fig. 2.13 Resonant tunneling diode (schematic)

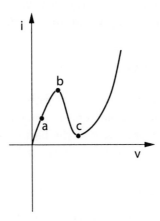

Fig. 2.14 vi-characteristic of a resonant tunneling diode; $v \geq 0$

The RTD is a nanoelectronic device based on tunnel currents, and as such only of minor importance in circuits based on single-electron electronics. Therefore only its operating principle is briefly outlined; for a more detailed introduction see for example [Goser *et al.* (2003)]. The nonlinear vi-characteristic, figure Fig. 2.14 shows a negative resistance part just as in the case of the tunnel diode. The mechanism responsible for this behavior is resonant tunneling.

In the quantum well between the two barriers a so-called *resonant* state is formed. The well is small, so distinct energy levels will be formed (section 4.5). In the well, electrons can only occupy one of these bound states. These bound states are, however, not true bound states as in the case of a quantum well with infinite thick barriers. Electrons can tunnel through one of the barriers to or escape from the well. As a consequence, the electrons will remain in the well for a certain time, and a quasi-bound or resonant state is formed. According to the Heisenberg uncertainty relation between time and energy (section 4.1), the energy in this state is spread into a range $\hbar/\tau$, where τ is the lifetime of an electron in the well before it escapes and $\hbar$ is Planck's constant divided by 2π. The electronic behavior of the RTD can now be visualized by figure Fig. 2.15.

The application of resonant tunneling diodes is in the combination with a transistor, e.g., a FET. The transistor can be used as a current source and the, dynamic system of the RTD has the two possible states. As such the combinations can be used for memory and for logic. In this way fast logic circuits have been realized, with speeds over ten gigahertz. Operation frequency has been projected upto and over 1 THz [Sze and Ng (2007)]

2.2   Tunneling Capacitor

In this section we turn to the experiments with the tunneling capacitor. In the experiments one or more capacitors are manufactured with a separation between the plates of the same size as the barrier in the tunnel diode; the separation, mostly by insulator material, is smaller than 10 nm.

2.2.1   *Tunneling between plates and the hot electron*

Almost immediately after the discovery of the tunnel diode, experiments were performed to find tunneling between two metal plates. It is probably the easiest way to perform a tunneling experiment. These experiments were conducted immediate after the discovery of the tunnel diode [Giaever

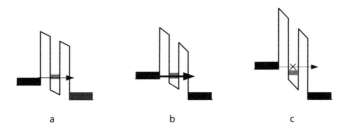

Fig. 2.15   Energy-band diagram and electron tunnel current in a resonant tunneling diode with increasing bias.

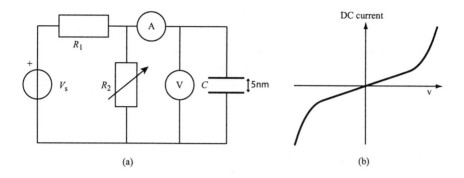

Fig. 2.16 Tunnel experiment and voltage-current characteristic; the characteristic is linear at low voltages.

(1960)].

An experiment is the following. Consider a simple DC circuit, consisting of two parallel plates connected to a battery. Include an ammeter, a voltmeter, and some resistors in the circuit so that we can measure the current through and the voltage over the capacitor. Figure 2.16(a) shows a schematic drawing of the circuit. Normally, we do not expect a DC current to flow between the capacitor plates; only a time-dependent current will be flow through the circuit until the steady-state is reached. If the capacitor plates are spaced very close together (on the order of 5 nm), it is found that a DC current is easily detected, and the voltage-current characteristic is as in figure Fig. 2.16(b). The DC current flowing through the thin isolator between the capacitor plates is a tunneling current.

In general this experiment can be conducted by evaporating a metal film onto a flat glass surface. Then expose it to a normal atmosphere. The metal surface will oxidize and forms a metal oxide that will act as an insulator. Evaporation of still another metal film will result in a M-I-M sandwich. Several metals will form a good, tough, insulating oxide. Relatively good to use are: Al, Cr, Ni, Mg, Nb, Ta, Sn, and Pb, in this order; more difficult to use are Cu, La, Co, V, and Bi; and Ag, Au, and In are impossible to use in this way [Giaever (1969)].

The linear behavior in the vi-measurements at low voltages give rise to the definition of a tunnel "resistance". Given the applied bias voltage, if the tunnel current is high this resistance is low; if the tunnel current is low this resistance is high. The tunnel current is expressed with a transmission probability, the probability for an electron to tunnel through the potential barrier. Tunneling will take place provided the tunneling electron can find an empty state on the other side with the same energy. If m is the electron mass, $\hbar$ Planck's constant divided by 2π, t the separation distance between the plates, and ϕ the metal-dielectric work function, this tunneling resistance for wide barriers at low voltages can be written as [Giaever (1969)]

$$R_t \approx \exp[(2t/\hbar)(2m\phi)^{1/2}] \approx \exp(t\phi^{1/2}). \tag{2.12}$$

In the last expression correct dimension are obtained by using t in angstroms and ϕ in electron volts. We see that the decrease in current with the thickness t of the barrier is exponential. Since the values in equation 2.12 are in the order of angstroms (one-tenth of a nanometer) a change of only 1 Å in the separation distance changes the resistance by an order of magnitude if the metal-dielectric work function is about 2 eV. The only reason that there can be an appreciable current flow through barriers of a few nanometers in thickness is the fact that the pre-exponential factor in the tunnel resistance is quite large.

Electrons that tunnel across the thin insulating film arrive in the second metal as hot electrons. When an electron tunneling from the Fermi level of the first metal enters the second metal, it is "hotter" by just about eV, where V is the voltage across the M-I-M sandwich. The question how far the electron can travel in the metal before it loses its excess energy is important for possible application in so-called metal-interface amplifiers or also called (tunnel-emitter) hot-electron transistors. Two types of collisions possible are electron-lattice and electron-electron. In metals, the latter is important because of the high free-electron concentration. In a collision process two electrons share the energy. Repeated collisions drop the energy of all hot electrons, and eventually, some distant away from the interface, they become part of the normal free-electron distribution.

The hot electron concentration drop approximately exponentially toward the interior of the metal. An **attenuation length** can be defined in the direction perpendicular to the interface, which may be smaller than the mean free path [Lindmayer and Wrigley (1965)]; for the definition and a discussion of the mean free path see section 5.1. The attenuation length is usually a few tens nanometers. If the metal is thinner than the attenuation length, hot electrons will appear on the open surface and "cold" emission can be obtained if their energy is greater than the work function. When the thin metal is followed by another tunneling interface, that is another small insulator layer, a tunnel emission amplifier or hot-electron transistor

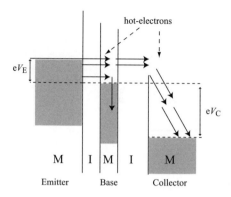

Fig. 2.17   Hot-electron transistor based on a M-I-M-I-M structure: energy band diagram (schematically)

can be obtained. Figure 2.17 shows schematically a MIMIM hot-electron transistor.

Over the years, hot-electron transistors (HETs) have been proposed replacing the third metal by a semiconductor or a heterojunction. Since a hot-electron, or more generally a hot carrier, has a higher energy and thus a higher velocity HETs are expected to have a higher intrinsic speed than bipolar transistors or FETs (field-effect transistors). More on these devices, which operation is based on currents (although individual single hot-electrons are essential) and thus fall outside the scope of this monograph can be found, for example, in [Sze and Ng (2007)] and refs. therein.

The linear behavior of the tunneling M-I-M structure, from now on called tunnel junction, is not very interesting for electronic applications. It provides circuits with functionality similar to pure resistive circuits. For more interesting applications a device with a nonlinear behavior is necessary. Such a nonlinear behavior was found by cooling down the tunneling M-I-M devices; a non-linear *vi*-characteristic was found when the metals became superconductive [Giaever (1960)]. This discovery led to the research on superconductive devices that resulted in the Josephson junction, SQUIDS (superconducting quantum interference devices), and RS-FQD (rapid single flux quantum devices). Together with the approach with the single-electron transistors, as is pursued in this text, the single flux quantum devices belong to the most interesting approaches to nanoelectronics. With RSFQ devices various logic circuits have been realized. Systems composed of these circuits allow a high-speed operation with clock

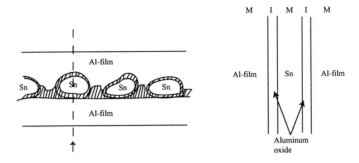

Fig. 2.18   Tunneling through particles of tin: model and schematic of the cross-section at the dashed line.

frequencies of upto 900 GHz. An introduction to these devices and their relation to circuit design can be found in [Goser *et al.* (2003)], for example. Although the superconductive devices have some attractive and interesting characteristics, their application in circuits and electronic systems is limited. This is mainly due to the necessity of the very low temperature ranges that are inherent to superconductivity.

In the search for more tunneling devices, new structures with nonlinear *vi*-characteristics were found. In studying superconductivity of small tin particles embedded in an oxide film, Giaever and Zeller found that around zero bias the tunnel junction shows a large peak in resistance *even* for non-superconducting particles [Giaever and Zeller (1968)]. Contact was made with the tin particles by tunneling through an oxide layer.

The experiment uses small Sn islands sandwiched between two aluminum layers, as shown in figure Fig. 2.18. Because the space between the particles is essentially filled with thick aluminum oxide, the tunnel current between the two aluminum layers flows mainly through the particles. The typical particle size was 7-11 nm. The structure is essentially a M-I-M-I-M structure, that is two tunnel junctions in series. The measurements are done at very low temperatures and a sufficiently high magnetic field was applied to avoid superconductive states. The resistance peak was found as a rising differential resistance at very low bias voltages as shown in figure Fig. 2.19. The resistance peak in these experiments is a first indication of the existence of what was a decade later called Coulomb blockade, and was explained in terms of a capacitor model of the Sn particles. The current towards the tunnel junctions consists of individual electrons; when a single electron tunnels onto or from the island the voltage across the tunneling

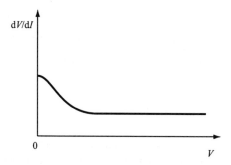

Fig. 2.19   Resistance peak in the tunneling current through the Sn particles.

junctions will change because the charge on these capacitors will change. The resistance peak indicates that, in *this* circuit, a minimum amount of energy is necessary to compensate for the increase in energy on the particle in the experiment.

2.2.2   *Tunnel capacitor and electron-box*

When in a M-I-M-I-M structure one of the insulator layers is thick, tunneling can only appear at the tunnel junction (the other M-I-M structure). These structures are know as electron-box structures. A first proposal of such a structure [Lambe and Jaklevic (1969)] consisted again of an array of metal droplets deposited, now, on a thick aluminum layer. The structure makes possible the study of microscopic capacitors of such size that the addition of a single electron causes a significant voltage change on the small capacitance of the tunnel junction. Figure 2.20 shows the structure and the equivalent circuit Lambe and Jaklevic proposed. Because the second capacitor is thick electrons can only tunnel on or from the island through the tunnel junction. The tunneling junction is represented by a tunnel resistance and a shunt capacitor. The capacitance formed by the thick oxide layer, the droplet, and the aluminum layer is not shunted by any appreciable leakage path. By raising the voltage the number of electrons on the droplet is expected to increase as a function of the applied voltage (in this case, any time an electron tunnels to the droplet the energy of the droplet is raised and requires more energy provided by an energy source, here the voltage source). The exact calculations will be discussed in chapter 9. Using, what they called a straightforward thermodynamic calculation, Lambe and Jaklevic predicted a staircase-like behavior, see figure Fig 2.21. It is

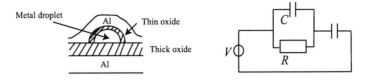

Fig. 2.20   Schematic structure of one element, and proposed equivalent circuit.

the property of the ever increasing number of electrons that can be stored on or retrieved from the droplet that coined the name electron box for this structure.

2.2.3   *Single-electron tunneling transistor and Coulomb blockade*

The next development was the observation of single-electron charging effects in macroscopic metallic islands still having small area tunnel junctions, instead of metal grains, by Fulton and Dolan. Of course such a development opens the possibility to integrate such devices in current microelectronic (or nanoelectronic) technologies. It also brings the description of the circuits within the domain of (linear) circuit theory. They observed charging effects in a configuration of three junctions of small area formed on a small common electrode [Fulton and Dolan (1987)].

The configuration has a low stray capacitance such that the capacitance charged and discharged by the tunneling electrodes is small and well defined. The tunnel junctions are modeled by a capacitor in parallel with a voltage controlled current source, representing a nonlinear resistor, as was

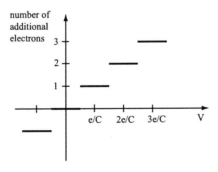

Fig. 2.21   Number of additional electrons as a function of the applied voltage; C is the capacitance of the nontunneling capacitor.

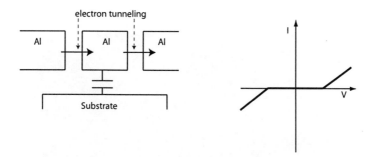

Fig. 2.22 Single-electron tunneling transistor structure, *vi*-charateristic schematically.

shown in figure Fig. 1.4. Because charge on the common electrode can be controlled the structure is called a single-electron tunneling transistor (SET transistor); charge variations on this common electrode control the tunneling through the tunneling junctions [Averin and Likharev (1986)]. The structure and the experimental characteristic, when the circuit was biased with a voltage source, are schematically shown in figure Fig. 2.22.

The voltage range in the nonlinear *vi*-characteristic in which no current flows through the tunnel junctions was called "Coulomb gap". Later on the common name for this behavior became "Coulomb blockade". The junctions were fabricated by use of a lift-off stencil formed through electron-beam lithography. The Al-Al tunnel junctions were formed using a multiple-angle deposition-oxidation-deposition cycle. A technique that was in use for many years. The junction resistances were $\approx 40\text{k}\Omega$; and the *vi*-curves were measured in the temperature range of 4.2-1.1 K. Examples are [Geerligs *et al.* (1990)] and [Heij *et al.* (2001)].

Nowadays, SET transistor structures can be manufactured almost routinely and many experiments have been done, mostly at low temperatures. The voltage controlling the charge on the common electrode, nowadays called gate electrode, V_g and the voltage of the bias source V_s are then varied. The characteristics are shown in three dimensions. Along the x-axis the gate voltage, on the y-axis the bias- or source voltage, and in different gray values the current through the tunnel junctions. An example is shown in figure Fig 2.23. The black diamond-shaped areas around $V_s = 0$ are the combinations of the gate voltage and bias voltage in which no tunneling is possible, that is, the circuit is in Coulomb blockade. In chapter 9 the SET-transistor characteristics and Coulomb blockade will be calculated fully.

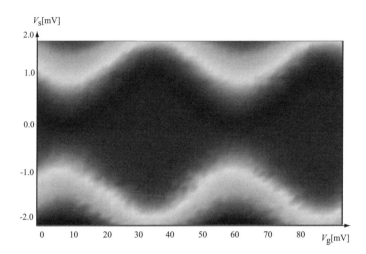

Fig. 2.23   Measurements of Coulomb blockade in the SET transistor.

2.2.4   *Array of tunnel junctions and Coulomb oscillations*

The discrete nature of charge can be observed directly in a single- or in an array of tunnel junctions by applying a direct current source. When the tunnel junction capacitance(s) is(are) excited by a DC current source, the capacitance will be charged until the voltage across the tunnel junction(s) will overcome the Coulomb blockade. As soon as this happens an electron will tunnel through the tunnel junction(s). By doing so the charge on the tunnel junction(s) will be inverted and the current source will start discharging and recharging the junction capacitance(s) again. Charging will continue until the Coulomb blockade can be overcome again, and the process will start again. In this way we will be able to observe voltage oscillations across the tunnel junction(s) with a frequency $f = I/e$; for the derivation of this formula see section 8.1. This phenomenon called Coulomb oscillations has been observed [Bylander *et al.* (2005)].

The experiment involved a superconducting array of 50 tunnel junctions excited by a bias dc current source. A magnetic field was applied in order to suppress superconducting particles (Cooper-pairs) to tunnel and, thus, to facilitate only single-electron tunneling. The measured power density spectrum clearly indicates a single frequency. This example shows that individual electrons can be observed, that the continuous current delivered by the source can be combined with the discrete character of electron tunneling, and that the tunnel junction array can be represented by capacitors as long as the Coulomb blockade holds. In this way the system can be an object of linear circuit analysis.

2.2.5 *Single-electron tunneling junction and Coulomb blockade*

For long time it had been thought that the property of Coulomb blockade was only due to the existence of an island in the structures. There exists some experiments that, at least, suggest that this may not be the case. To feed this discussion some researchers try to observe single electron charging effects in single-junctions excited by a voltage source (note that a minimum of two junctions, or a junction and a capacitor, are necessary to create an island).

Already in the first SET transistor experiment [Fulton and Dolan (1987)] there was mention of a possible Coulomb gap (Coulomb blockade) in "one of the three junctions of a sample of small capacitance". Direct measurements on a solitary voltage biased tunnel junction have been done by [Kauppinen and Pekola (1996)]. They found that even for the single tunnel junction "based on clear-cut results in the high temperature regime" Coulomb blockade is observed. Experiments like these support the developed circuit theory for single-electron tunneling circuits in this book; in the circuit theory the Coulomb blockade appears as a property of the tunnel junction.

At the end of this chapter measurement results are shown of a so-called single-electron pump, see figures Fig. 2.24 and Fig. 2.25. The electron pump circuit still belongs to the most advanced circuits with SET technology in 2008. It is a circuit able to pump around single electrons in a controlled manner in order to define a precise current. The different triangles show stable states in which additional electrons are present on different islands.

Problems and Exercises

2.1 Calculate the space-charge width in an abrupt Ge $p - n$ junction for $N_a = 10^{23} \mathrm{m}^{-3}$, $N_d = 10^{22} \mathrm{m}^{-3}$, and $\epsilon_r = 16$.

2.2 Show that for small values of the bias voltage the tunnel current (equation 2.9) is directly proportional to the applied voltage. (hint: expand the Fermi-Dirac function to lowest order in a Taylor series.)

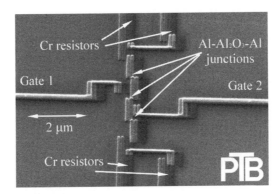

Fig. 2.24 A single-electron pump. The single-electron pump belongs to the most advanced SET circuits. The device was manufactured by S.V. Lotkov and A.B. Zorin from PTB (Physikalisch Technische Bundesanstalt) [Lotkhov *et al.* (2001)] and measured at the Dutch national measurement standards laboratory - NMI Van Swinden Lab.

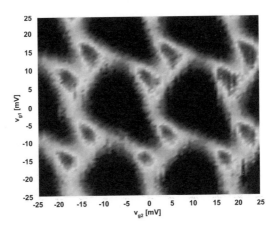

Fig. 2.25 Measurements on a single-electron pump.

2.3 Consider direct band-to-band tunneling in a tunnel device. Although no energy is dissipated during tunneling, energy is dissipated in the tunnel diode device. Propose dissipation mechanisms, and explain them.

2.4 Explain the working of a hot-electron transistor. Look for applications in existing literature. Why is it called a transistor?

2.5 Is the Coulomb blockade phenomenon found experimentally in a single SET junction excited by a voltage source? Consult literature.

Chapter 3

Current in Electrodynamics and Circuit Theory

This chapter provides a brief review of the modeling of currents in classical physics. A good understanding of currents is essential in any attempt to make the connection between the new nanoelectronic devices and a suitable circuit description. In the sections, the various currents are defined precisely and approximations are made explicitly. Although in the theory currents express the conservation of charge, the conservation of energy and its consequences for the theory of circuits is discussed too.

Electric currents are electrons or other charges in motion with a net drift or flow. In electronic devices we can consider currents that may consist of movement of electrons due to an electric field called drift current, movement of electrons due to a gradient in the electron concentration called diffusion current, or the movement of electrons due to tunneling called tunnel current. In this chapter attention is focussed on the drift of electrons in the electromagnetic field. Classical the electrons are seen as particles and current can be defined as the movement of charge particles. The electromagnetic field is described in accordance with Maxwell's equations.

3.1 Charges in Electrodynamics

Any current consists of moving charged particles. In general, the charge particles are considered to be very small compared with their relative separation, and their electric charge is concentrated at points (point charges). It is, however, often more convenient to think of charge distributions as being continuous instead of a consisting of a series of point charges. We follow the treatment in [de Hoop (1975)]. A **fundamental volume** V_{gr} is defined with the properties that it is large enough to show macroscopic behavior, but small when compared with the total volume. V_{gr} is the smallest

Table 3.1　Important quantities and symbols used in this chapter

Quantity	Symbol	Unit	Symbol
length	l	meter	m
area	A	square meter	m^2
volume	V	cubic meter	m^3
concentration	N	per cubic meter	m^{-3}
time	t	second	s
energy	W	joule	J
power	P	watt	W
Poynting vector	$\mathbf{S}$	watt per cubic meter	W/m^3
force	$\mathbf{F}$	newton	N
volume density of the force	$\mathbf{f}$	newton per cubic meter	N/m^3
velocity	$\mathbf{v}$	meter per second	m/s
charge	Q	coulomb	C
charge density	ρ	coulomb per cubic meter	C/m^3
current	I	ampère	A
current density	$\mathbf{J}$	ampère per square meter	A/m^2
voltage difference	V	volt	V
electric field strength	$\mathbf{E}$	volt per meter	V/m
magnetic field strength	$\mathbf{H}$	ampère per meter	A/m
permeability of the vacuum	μ_0	henry per meter	H/m
permittivity of the vacuum	ϵ_0	farad per meter	F/m
conductivity	σ	siemens per meter	S/m
resistance	R	ohm	Ω
heat per unit volume	Q	joule	J
self-inductance	L	weber per ampere	W/A
fundamental charge	e	coulomb	C

possible grain in which the collective behavior of charge particles can be described by macroscopic physics laws.

Suppose Δn is the number of electrons (or more general charged particles) in volume ΔV. Then the **concentration** $N = N(\mathbf{r}, t)$ of electrons in V can be defined as

$$N \overset{\text{def}}{=} \lim_{\Delta V \to V_{\text{gr}}} \Delta n / \Delta V, \qquad (3.1)$$

in which $\mathbf{r}$ is a point within V_{gr}. $N = N(\mathbf{r}, t)$ is a continuous function.

Using the above definition the **total number of particles** $n = n(t)$ in a volume V at time t is

$$n \overset{\text{def}}{=} \int_V N \, dV, \qquad (3.2)$$

in which $dV = dx dy dz$.

In case of a current the particles are charged and move. To describe this movement the **average velocity** $<\mathbf{v}>$ of the particles numbered k $(k = 1, 2, ..., \Delta n)$ in ΔV is introduced:

$$<\mathbf{v}> = (\Delta n)^{-1} \sum_{k=1}^{\Delta n} \mathbf{v}_k. \tag{3.3}$$

A continuous quantity $\mathbf{v}_{\mathrm{conv}} = \mathbf{v}_{\mathrm{conv}}(\mathbf{r}, t)$ called the **convection velocity** of the particle can be obtained by taking the limit

$$\mathbf{v}_{\mathrm{conv}} \stackrel{\mathrm{def}}{=} \lim_{\Delta V \to V_{\mathrm{gr}}} <\mathbf{v}>. \tag{3.4}$$

The macroscopic concepts (electric) charge density $\rho = \rho(\mathbf{r}, t)$ and (electric) current density $\mathbf{J} = \mathbf{J}(\mathbf{r}, t)$ can be introduced, now, by considering a swarm of charged particles concentrated within V_{gr}. The **charge density** $\rho = \rho(\mathbf{r}, t)$ is now a continuous function defined as

$$\rho \stackrel{\mathrm{def}}{=} \lim_{\Delta V \to V_{\mathrm{gr}}} (\Delta V)^{-1} \sum_{k=1}^{\Delta N} q_k \tag{3.5}$$

and the **current density $\mathbf{J} = \mathbf{J}(\mathbf{r}, t)$** as

$$\mathbf{J} \stackrel{\mathrm{def}}{=} \lim_{\Delta V \to V_{\mathrm{gr}}} (\Delta V)^{-1} \sum_{k=1}^{\Delta N} q_k \mathbf{v}_k. \tag{3.6}$$

In the equations q_k is the charge of particle k and $\mathbf{v}_k$ its velocity. The in this way defined charge density $\rho = \rho(\mathbf{r}, t)$ and current density $\mathbf{J} = \mathbf{J}(\mathbf{r}, t)$ are continuous functions. Using equation 3.1 and averages, like equation 3.3, equations 3.5 and 3.6 become:

$$\rho = N \lim_{\Delta V \to V_{\mathrm{gr}}} <q> \tag{3.7}$$

and

$$\mathbf{J} = N \lim_{\Delta V \to V_{\mathrm{gr}}} <q\mathbf{v}>, \tag{3.8}$$

N again being the concentration of particles in the fundamental volume.

For a current of electrons the charge q is the elementary electron charge e $(e = 1.60 \times 10^{-19}$ C$)$ and is the same for all electrons, the charge density becomes

$$\rho = Ne \tag{3.9}$$

and the electric current density becomes

$$\mathbf{J} = Ne\mathbf{v}_{\text{conv}}. \tag{3.10}$$

The **total charge** $Q = Q(t)$ in V at time t can now be defined as

$$Q \overset{\text{def}}{=} \int_V \rho \, dV, \tag{3.11}$$

and the **total current** $I = I(\mathbf{r}, t)$ is defined as a flux of the electric current density through a surface S pointed outwards:

$$I \overset{\text{def}}{=} \oint_S \mathbf{n} \cdot \mathbf{J} \, dA, \tag{3.12}$$

where $\mathbf{n}$ is the **unit vector normal to** S.

3.2   Conservation of Charge and Continuity Equation

To find a relation between $\rho(\mathbf{r}, t)$ and $\mathbf{J}(\mathbf{r}, t)$ we consider a fixed volume V enclosed by an area S in which an amount charge Q is located, and consider a current directed outwards this volume. We know that charged must be conserved as required by the physic's law of conservation of charge. The amount of charge flowing out of the volume V must be a result of a decrease of charge inside the volume. The amount of charge inside the volume at time t is $Q(t)$,

$$Q(t) = \int_V \rho(\mathbf{r}, t) dV,$$

and at time $t + \Delta t$ the amount of charge is $Q(t + \Delta t)$,

$$Q(t + \Delta t) = \int_V \rho(\mathbf{r}, t + \Delta t) dV.$$

The amount of charge flowing out of the closed surface is

$$\oint_S \mathbf{n} \cdot \mathbf{J} \Delta t dA.$$

Conservation of charge is, now, expressed by

$$Q(t + \Delta t) - Q(t) = -\oint_S \mathbf{n} \cdot \mathbf{J} \Delta t dA.$$

Dividing by Δt, taking the limit $\Delta t \to 0$ and writing the charge as the integral of the charge density, we obtain:

$$\int_V \frac{d\rho}{dt}\, dV + \oint_S \mathbf{n} \cdot \mathbf{J}\, dA = 0 \tag{3.13}$$

or, using Gauss's divergence theorem,

$$\frac{\partial \rho}{\partial t} + \operatorname{div} \mathbf{J} = 0. \tag{3.14}$$

Equation 3.13 expresses the **conservation of electrical charge**, it holds for every closed surface S. Equation 3.14 is known as the **continuity equation for electrical charge**. Equation 3.13 is in agreement with the definition of current as the charge passed through the total cross-sectional area of a conductor per unit time

$$I \stackrel{\text{def}}{=} \frac{dQ}{dt}. \tag{3.15}$$

Example 3.1 The expression for the charge $Q = Q(t)$ inside the surface S in terms of the current $I = I(t)$ flowing outwards through the surface in the time interval $t_0 > t > \infty$.

∇
Starting from the definition of the current equation 3.12 or just integrating equation 3.15 we obtain:

$$Q(t) = Q(t_0) - \int_{t_o}^{t} I(\tau)d\tau.$$

$\triangle$

3.3 Electromagnetics' Field Equations in Vacuum

Electrical particles attract or repel each other. These forces depend on their positions and their velocities, and are a consequence of the fact that electrical charged particles always give rise to the presence of electromagnetic fields — described by electromagnetic field vectors.

By measuring the force $\mathbf{F}$ experienced by a charge, a relation may be deduced. $\mathbf{F}$ is proportional to its electric charge q; $\mathbf{F}$ is proportional to its velocity **v** measured by an observer in the direction perpendicular to **v**; **F** has a term that is independent of **v**. The result is that

$$\mathbf{F} = q\mathbf{E} + q\mathbf{v} \times \mu_0\mathbf{H}, \tag{3.16}$$

in which μ_0 is the permeability of the vacuum; $\mathbf{E} = \mathbf{E}(\mathbf{r}, t)$ is called the **electric field strength** and $\mathbf{H} = \mathbf{H}(\mathbf{r}, t)$ the **magnetic field strength** at the location of the charge, and both are produced by all the other charges that are present. The force $\mathbf{F} = \mathbf{F}(\mathbf{r}, t)$ is called the **Lorentz force**.

The description above presents a conceptual problem: although the electromagnetic field can be described mathematically without the need for any test charges, it cannot be measured without them. On the other hand, the test charge itself gives rise to an electromagnetic field. However, it is impossible to measure an external field with a test charge and, at the same time, to determine the field due to this charge.

We can determine the electromagnetic field vectors in vacuum. For this case, the force on a single charged particle is used to define the electromagnetic field vectors. That is, the electrical charge is used as a measuring object to find the values of the electromagnetic field vectors. Because this measuring object cannot be placed in matter we can only define electromagnetic field vectors in vacuum—that is, in the absence of other charges or currents—in this way.

Besides this, it is assumed that the strength of the fields caused by the measuring charge can be neglected in measuring the intensity of the electric field and the intensity of the magnetic field of the original electromagnetic field present in the vacuum The fields can be found using equation 3.16.

In vacuum, various experiments on the relations between electric and magnetic fields can be summarized by **Maxwell's equations in vacuum**:

$$\operatorname{curl}\mathbf{H} - \epsilon_0\frac{\partial\mathbf{E}}{\partial t} = 0, \tag{3.17}$$

$$\operatorname{curl}\mathbf{E} + \mu_0\frac{\partial\mathbf{H}}{\partial t} = 0, \tag{3.18}$$

$$\operatorname{div}\mathbf{E} = 0, \tag{3.19}$$

$$\operatorname{div}\mathbf{H} = 0. \tag{3.20}$$

In the equations is

$$\epsilon_0 \overset{\mathrm{def}}{=} (\mu_0 c^2)^{-1}, \tag{3.21}$$

ϵ_0 is the permittivity of vacuum and c the speed of light in vacuum.

3.4 Equations in the Presence of Charges and Currents

Now, we will go to the description of the properties of the electromagnetic field vectors in an area in which charged particles or currents are present. This has to be done carefully, because the electromagnetic field vectors can not be measured directly in such area.

A suitable starting point for the description of the electromagnetic field vectors is Gauss's law. Therefore a closed surface S is chosen such that S is in vacuum and all charges are inside. Based on experiments, Gauss's law may then be stated: the net electric flux through a closed surface S equals the total net charge inside Q the surface divided by the permittivity of the vacuum

$$\oint_S \mathbf{n} \cdot \mathbf{E} \, dA = \frac{Q}{\epsilon_0}, \tag{3.22}$$

If $\rho = \rho(\mathbf{r}, t)$ is the charge density inside the S, we write

$$Q = \int_V \rho \, dV$$

Assuming that, in these cases, $\mathbf{E}$ indexelectric field is continuously differentiable Gauss's theorem may be applied. This results in:

$$\int_V \operatorname{div} \mathbf{E} \, dV = \epsilon_0^{-1} \int_V \rho \, dV, \tag{3.23}$$

or

$$\operatorname{div} \mathbf{E} = \frac{\rho}{\epsilon_0}. \tag{3.24}$$

This is **Gauss's law of the electric field**.

Experiments measuring $\mathbf{H}$ on the surface S reveal:

$$\oint_S \mathbf{n} \cdot \mathbf{H} \, dA = 0, \tag{3.25}$$

or

$$\operatorname{div} \mathbf{H} = 0. \tag{3.26}$$

This is **Gauss's law for magnetism**, expressing that magnetic monopoles do not exist.

Experiments show that moving charged particles and currents generate a magnetic field. The correct relation between the current density and the

Table 3.2　Electromagnetic field vectors inside matter

Quantity	Symbol	Unit	Symbol
electric polarization	P	coulomb per cubic meter	C/m^2
magnetization	M	ampère per meter	A/m
electric flux density	D	coulomb per cubic meter	C/m^2
magetic flux density	B	tesla	T

magnetic field strength was found by Maxwell, Maxwell's law for electromagnetic fields,

$$\operatorname{curl} \mathbf{H} = \mathbf{J} + \epsilon_0 \frac{\partial \mathbf{E}}{\partial t}, \tag{3.27}$$

which holds for all currents. Still holds

$$\operatorname{curl} \mathbf{E} = -\mu_0 \frac{\partial \mathbf{H}}{\partial t}, \tag{3.28}$$

known as **Faraday's law of induction.**

Equations 3.24, 3.26, 3.27, and 3.28 are the **Maxwell equations in vacuum in the presence of charges and currents** as put forward by Lorentz in 1892 in his electron theory (in modern notation); the charges or currents are always described inside a closed surface, while the electromagnetic field vectors are measured along or outside the closed surface.

We conclude this section by making two further points regarding the Maxwell equations. First, inside conductors—or more general inside matter—electromagnetic field vectors also exist. Because they cannot be measured directly, they are introduced axiomatically with the vectors: **P** the electrical polarization, **M** the magnetization, **D** the electric flux density or displacement, and **B** the magnetic flux density or magnetic induction, see also table 3.2. Second, for the accurate description of currents the Maxwell equations are completed with the continuity equation, equation 3.14, expressing the conservation of charge.

Many books on electrodynamics prefer to describe the magnetic field by means of only one quantity **B**, the magnetic flux density. The difference between both quantities becomes only important in matter when there is both an **amperian current** and a **conduction current**. Amperian currents are microscopic currents within molecules and atoms; its the hypothetical currents that causes magnetization in a magnetic material. To describe conduction, however, we only need the conduction current due to freely moving conduction electrons. In contrast to the magnetic flux density **B**,

Table 3.3   Electromagnetic field vector equations

Lorentz force equation:
$$\mathbf{F} = q(\mathbf{E} + \mathbf{v} \times \mu_0\mathbf{H}) \qquad\qquad \mathbf{F} = q(\mathbf{E} + \mathbf{v} \times \mathbf{B})$$

Maxwell equations:

$$\text{div}\,\mathbf{E} = \rho/\epsilon_0 \qquad\qquad \text{div}\,\mathbf{E} = \rho/\epsilon_0$$
$$\text{curl}\,\mathbf{E} = -\mu_0\,\partial\mathbf{H}/\partial t \qquad \text{curl}\,\mathbf{E} = -\partial\mathbf{B}/\partial t$$
$$\text{div}\,\mathbf{H} = 0 \qquad\qquad \text{div}\,\mathbf{B} = 0$$
$$\text{curl}\,\mathbf{H} = \epsilon_0\,\partial\mathbf{E}/\partial t + \mathbf{J} \qquad c^2\,\text{curl}\,\mathbf{B} = \partial\mathbf{E}/\partial t + \mathbf{J}/\epsilon_0$$

definition vector potential:
$$\mathbf{H} = \mu_0^{-1}\text{curl}\,\mathbf{A} \qquad\qquad \mathbf{B} = \text{curl}\,\mathbf{A}$$

energy equation:
$$W = \int(\tfrac{1}{2}\epsilon_0\mathbf{E}\cdot\mathbf{E} + \tfrac{1}{2}\mu_0\mathbf{H}\cdot\mathbf{H})\mathrm{d}V \qquad W = \int(\tfrac{1}{2}\epsilon_0\mathbf{E}\cdot\mathbf{E} + \tfrac{1}{2}\epsilon_0 c^2\mathbf{B}\cdot\mathbf{B})\mathrm{d}V$$

---

the magnetic field strength $\mathbf{H}$ only depends on the conduction current. Because we are going to formulate a circuit theory, in which only conduction currents (or more specific, conduction electrons) will play a role, it is quite natural to use $\mathbf{H}$. The table 3.3 shows the corresponding formula.

3.5   Conservation of Energy and Poynting's Theorem

The electromagnetic field will exert a Lorentz force given by equation 3.16 on a charged particle. The **work** done by the applied force on the particle when it moves through a displacement ΔW is defined to be $\Delta W = \mathbf{F}\cdot\Delta\mathbf{r}$ and the rate at which the work is done is the **power** P

$$P = \frac{\mathrm{d}W}{\mathrm{d}t} = \mathbf{F}\cdot\mathbf{v} = q\mathbf{E}\cdot\mathbf{v}, \qquad (3.29)$$

because $\mathbf{v}\cdot(\mathbf{v}\times\mathbf{B}) = 0$; the magnetic field does no work on the charged particle. If we interpret $\mathbf{v}$ as the convection velocity and using $\mathbf{J} = \rho\mathbf{v}_{\text{conv}}$ we define

$$\frac{\mathrm{d}w}{\mathrm{d}t} = \mathbf{E}\cdot\mathbf{J}, \qquad (3.30)$$

in which w is the work per unit volume.

The expression obtained is for the energy associated with the motion of charged particles.

What we would like to have is an expression for the energy that applies to a general distribution of charges and currents $(\rho, \mathbf{J})$. We find

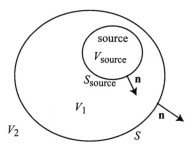

Fig. 3.1 Exchange of electromagnetic energy.

this expression by using the Maxwell equations in vacuum in the presence of currents equations 3.24 and 3.26. Multiplying equation 3.27 by $\mathbf{E}$ and equation 3.28 by $\mathbf{H}$ and subtracting one from the other, we get using $\text{div}(\mathbf{E} \times \mathbf{H}) = -\mathbf{E} \cdot \text{curl}\,\mathbf{H} + \mathbf{H} \cdot \text{curl}\,\mathbf{E}$:

$$\mathbf{E} \cdot \mathbf{J} = -\epsilon_0 \mathbf{E} \cdot \frac{\partial \mathbf{E}}{\partial t} - \mu_0 \mathbf{H} \cdot \frac{\partial \mathbf{H}}{\partial t} - \text{div}\,(\mathbf{E} \times \mathbf{H}). \tag{3.31}$$

Using $\mathbf{E} \cdot \partial\mathbf{E}/\partial t = 1/2\partial(\mathbf{E} \cdot \mathbf{E})/\partial t$ and $\mathbf{H} \cdot \partial\mathbf{H}/\partial t = 1/2\partial(\mathbf{H} \cdot \mathbf{H})/\partial t$ we find

$$\mathbf{E} \cdot \mathbf{J} + \frac{\partial}{\partial t}(\frac{1}{2}\epsilon_0 \mathbf{E} \cdot \mathbf{E} + \frac{1}{2}\mu_0 \mathbf{H} \cdot \mathbf{H}) + \text{div}\,(\mathbf{E} \times \mathbf{H}) = 0. \tag{3.32}$$

The standard interpretation of this equation is found after integration of equation 3.32 over a closed area V_1 in which a source area—an area within which electromagnetic energy is generated—is excluded, see figure Fig. 3.1. The result of the integration is

$$\int_{V_1} \mathbf{E} \cdot \mathbf{J}\,dV + \frac{\partial}{\partial t}\int_{V_1}(\frac{1}{2}\epsilon_0 \mathbf{E} \cdot \mathbf{E} + \frac{1}{2}\mu_0 \mathbf{H} \cdot \mathbf{H})dV +$$

$$+ \oint_S \mathbf{n} \cdot (\mathbf{E} \times \mathbf{H})dA = \oint_{S_{\text{source}}} \mathbf{n} \cdot (\mathbf{E} \times \mathbf{H})dA, \tag{3.33}$$

where

$$\int_{V_1} \text{div}\,(\mathbf{E} \times \mathbf{H})dV = \oint_S \mathbf{n} \cdot (\mathbf{E} \times \mathbf{H})dA - \oint_{S_{\text{source}}} \mathbf{n} \cdot (\mathbf{E} \times \mathbf{H})dA \tag{3.34}$$

was used; equation 3.33 is known as **Poynting's theorem**. And the vector $\mathbf{S}$

$$\mathbf{S} = \mathbf{E} \times \mathbf{H} \tag{3.35}$$

is called the **Poynting vector**.

The physical interpretation of equation 3.33 is the following. First note that all of the terms (integrals) in the expression have units of energy per unit time [J/s] or power [W]; therefore, this relation is customarily viewed as a statement of **conservation of energy**. We examine all the terms in the theorem. The first term represents the rate at which energy is exchanged between the electromagnetic field and the mechanical motion of the charge within the volume V_1, see equation 3.30. The second term represent the electromagnetic energy stored in the electric field and the electromagnetic energy stored in the magnetic field in the volume V_1. The third term

$$P \stackrel{\text{def}}{=} \oint_S \mathbf{n} \cdot (\mathbf{E} \times \mathbf{H}) \mathrm{d}A \qquad (3.36)$$

is interpreted as the instantaneous power transported by the electromagnetic field from the volume V_1 to the volume V_2 through the surface S. In the same way the last term

$$P_{\text{source}} \stackrel{\text{def}}{=} \oint_{S_{\text{source}}} \mathbf{n} \cdot (\mathbf{E} \times \mathbf{H}) \mathrm{d}A \qquad (3.37)$$

can be interpreted as the instantaneous power provided by the energy source through S_{source} to the volume V_1.

Equation 3.33 can, now, be written as:

$$\frac{\mathrm{d}P_{\text{mech}}}{\mathrm{d}t} + \frac{\mathrm{d}W_{\text{em}}}{\mathrm{d}t} + P = P_{\text{source}} \qquad (3.38)$$

explicitly showing the conservation of energy.

Example 3.2 The Poynting vector $\mathbf{S} = \mathbf{E} \times \mathbf{H}$ applied to a conductor (wire) having a resistance when it is carrying a constant current.

∇

Since the conductor has resistance there is an electric field inside the conductor driving the current, and a voltage difference across it. Due to the continuity of the tangential component of the electric field there is also an electric field just outside the conductor, parallel to the surface. Due to equation 3.27 there is a magnetic field which goes around the conductor because of the current. The fields $\mathbf{E}$ and $\mathbf{H}$ are at right angles; therefore the Poynting vector is directed radially inward. There is a flow of energy into the conductor all around. This is, of course, equal to the energy being lost in the conductor in the form of heat. Instead of getting the energy from electrons being pushed along the conductor, the electrons are getting their energy from the field outside.

$\triangle$

3.6   Steady-State and Constant Currents

We have now gathered enough knowledge for describing currents. The first currents to be considered will be constant currents that belong to the class of the so-called **steady-state currents**. This *is* an approximation. It refers to a special kind of dynamic situation with currents consisting of *large numbers* of conduction electrons in motion that can be approximated by a *steady* flow of electrons. In this approximation the charge density $\rho(\mathbf{x}, t)$ does not change with time, that is, $\partial\rho/\partial t$ is zero. Consequently, $\partial\mathbf{E}/\partial t$ will be zero too.

Applying $\partial\rho/\partial t = 0$ to the continuity equation 3.14 and to Maxwell's equation for the magnetic field equation 3.27, we obtain for *steady-state currents*:

$$\operatorname{div} \mathbf{J} = 0 \qquad (3.39)$$

and

$$\operatorname{curl} \mathbf{H} = \mathbf{J}. \qquad (3.40)$$

Example 3.3 Show that the requirement $\partial\mathbf{E}/\partial t = 0$ and Maxwell's equation for the magnetic field

$$\operatorname{curl} \mathbf{H} = \mathbf{J} + \epsilon_0 \frac{\partial \mathbf{E}}{\partial t}$$

also leads to

$$\operatorname{div} \mathbf{J} = 0$$

∇

Starting from Maxwell's equation for the magnetic field and taking the divergence of both sides we obtain:

$$\operatorname{div}(\operatorname{curl} \mathbf{H}) = \operatorname{div} \mathbf{J}.$$

Because

$$\operatorname{div}(\operatorname{curl} \mathbf{h}) = 0$$

for any vector $\mathbf{h}$, we find:

$$\operatorname{div} \mathbf{J} = 0.$$

$\triangle$

3.6.1 *Kirchhoff's current law*

Equation 3.39, $\operatorname{div} \mathbf{J} = 0$, for steady-state currents has some important consequences. In particular, it leads to Kirchhoff's current law. Consider therefore the integral form of equation 3.39:

$$\oint_S \mathbf{n} \cdot \mathbf{J} \, \mathrm{d}A = 0. \tag{3.41}$$

Suppose we have a node in which several currents come together through N electrodes. Choose S to be the surface of the node. On the *isolated* parts of S holds:

$$\mathbf{n} \cdot \mathbf{J} = 0. \tag{3.42}$$

Through the N electrodes current can flow towards and from the node. Applying equation 3.41 on these electrodes immediately leads to:

$$\sum_{n=1}^{N} I_n = 0. \tag{3.43}$$

Equation 3.43 is known as **Kirchhoff's current law** (KCL).

3.6.2 *Vector potential* **A**

In 1820 in a famous experiment, Oersted discovered that a steady-state current produced a magnetic field

$$\operatorname{curl} \mathbf{H} = \mathbf{J}.$$

This **H** field can be calculated with the help of the so-called vector potential **A**. Gauss's law for magnetism

$$\operatorname{div} \mathbf{H} = 0$$

is automatically satisfied by setting

$$\mathbf{H} = \mu_0^{-1} \operatorname{curl} \mathbf{A}.$$

A is called the **vector potential** and is arbitrary to within a gradient:

$$\mathbf{A}' = \mathbf{A} + \operatorname{grad} \phi,$$

with ϕ an arbitrary scalar function; $\mathbf{H} = \mu_0^{-1} \operatorname{curl} \mathbf{A}'$ is also true.

Because of the vector identity

$$\operatorname{curl}\operatorname{curl}\mathbf{a} = \operatorname{grad}\operatorname{div}\mathbf{a} - \nabla^2\mathbf{a}, \tag{3.44}$$

the vector potential must satify the condition

$$\operatorname{grad}\operatorname{div}\mathbf{A} - \nabla^2\mathbf{A} = \mu_0\mathbf{J}. \tag{3.45}$$

(∇^2 is the Laplacian operator.) Equation 3.45 is satisfied if the following conditions are fulfilled together:

$$\operatorname{div}\mathbf{A} = 0 \quad \text{and} \quad \nabla^2\mathbf{A} = -\mu_0\mathbf{J}. \tag{3.46}$$

The choice div $\mathbf{A} = 0$ is used to determine the scalar function ϕ and is called the Coulomb gauge. The second part can be used to find an expression for the vector potential $\mathbf{A}$:

$$\mathbf{A}_p = \frac{\mu_0}{4\pi} \int \frac{\mathbf{J}_{p'}\,\mathrm{d}V_{p'}}{r_{pp'}}, \tag{3.47}$$

see for example [Feynman *et al.* (1964)].

3.6.3   *Ohm's law*

Other consequences are found for constant currents, in which case the magnetic field will be static too. In general, **steady-state currents** are define by the requirement that *all* fields are time independent, that is, $\partial\rho/\partial t \overset{\text{def}}{=} 0$, $\partial\mathbf{E}/\partial t \overset{\text{def}}{=} 0$, and $\partial\mathbf{H}/\partial t \overset{\text{def}}{=} 0$. The Maxwell equations will become

$$\operatorname{curl}\mathbf{H} = \mathbf{J}, \qquad \operatorname{div}\mathbf{H} = 0 \tag{3.48}$$

and

$$\operatorname{curl}\mathbf{E} = 0, \qquad \operatorname{div}\mathbf{E} = \frac{\rho}{\epsilon_0}. \tag{3.49}$$

The expression curl $\mathbf{E} = 0$ shows that in case of steady-state currents one of the Maxwell equations reduces to the same expression as the equation expressing energy conservation in a conservative field:

$$\oint \mathbf{E}\cdot\mathbf{ds} = 0 = e\oint \mathbf{E}\cdot\mathbf{ds} = \oint \mathbf{F}\cdot\mathbf{ds}.$$

It shows no net work can either be gained nor lost when a charged particle traverses a closed path in the electric field in case of steady-state currents. An alternative expression for this is

$$\mathbf{E} = -\mathrm{grad}\, V, \tag{3.50}$$

that is, a potential difference $V = u_1 - u_2$ can be defined between two points in the electric field (u_1 and u_2 are called the potential at point 1 and point 2, respectively), which value does not depend on the path taken.

Now, the resistance can be introduced by using the experimental fact that when a current flows in a material, the following phenomenological relation exists between the vectors $\mathbf{J}$ and $\mathbf{E}$. If σ is the **conductivity** of the material used then

$$\mathbf{J} = \sigma \mathbf{E}. \tag{3.51}$$

The above equation is called **Ohm's law,** and holds for those values of the electric field for which the relation between $\mathbf{J}$ and $\mathbf{E}$ is linear.

What happens in a homogeneous conductor, having a resistance, when it is carrying a current? We know that the electromagnetic energy to maintain the current is supplied by the source (source area). And, since the conductor has resistance, there is an electric field in the conductor, causing the current; without this field the electrons will not move other than in arbitrary directions due to temperature. There is also an electric field just outside the resistive conductor—the field we can measure.

For a homogeneous conductor, σ is independent of position, constant cross-sectional area A and length l using Ohm's law we obtain

$$u_1 - u_2 = l|\mathbf{E}| \tag{3.52}$$

$$I = A|\mathbf{J}| \tag{3.53}$$

so that

$$u_1 - u_2 = V = RI \tag{3.54}$$

where $R = l/(\sigma A)$. R is the resistance and depends upon the physical parameters of the conductor. Equation 3.54 is also valid for homogeneous conductors of arbitrary shape [Pauli (1973)] and is called **Ohm's law for the resistor.**

3.6.4 *Kirchhoff's voltage law*

For the last consequence of steady state currents treated here, consider energy conservation in case of a circuit consisting of an energy source delivering a constant current and a conductor with a resistance R. In steady-state we obtain from equation 3.38 that $|P_{\text{diss}}| = |P_{\text{source}}|$, if we do not allow energy to leave the circuit (remember that the fields are static and thus radiation cannot be included), that is, the electromagnetic energy delivered by the source is converted totally into heat in the conductor. By considering the work per unit time done by the force on the electrons, $\mathbf{F} = e\mathbf{E}$, it was found that the heat developed per unit volume per unit time is, see equation 3.30,

$$Q = \mathbf{J} \cdot \mathbf{E} \tag{3.55}$$

and is called **Joule heat**. The total heat produced per second in a homogeneous conductor of length l and cross-sectional area A is

$$\overline{Q} = \mathbf{J} \cdot \mathbf{E}Al = I(u_1 - u_2) = I^2R = P_{\text{diss}}. \tag{3.56}$$

On the other hand the power delivered by the source area:

$$P_{\text{source}} = \oint_S \mathbf{n} \cdot (\mathbf{E} \times \mathbf{H})\mathrm{d}A$$

can be found to be equal to [de Hoop (1975)]

$$P_{\text{source}} = \int_V \mathbf{E} \cdot \mathbf{J}\mathrm{d}V, \tag{3.57}$$

using $-\text{div}(\mathbf{E} \times \mathbf{H}) = \mathbf{E} \cdot \text{curl}\,\mathbf{H} - \mathbf{H} \cdot \text{curl}\,\mathbf{E}$, $\text{curl}\,\mathbf{E} = 0$, and $\text{curl}\,\mathbf{H} = \mathbf{J}$. Consequently,

$$P_{\text{source}} = -VI, \tag{3.58}$$

with V the potential difference across the source. This can be seen by noting that

$$\int_V \mathbf{E} \cdot \mathbf{J}\mathrm{d}V = -\int_V \text{grad}\,V \cdot \mathbf{J}\mathrm{d}V.$$

Using

$$\text{div}(V\mathbf{J}) = V\text{div}\,\mathbf{J} + \mathbf{J} \cdot \text{grad}\,V$$

one finds

$$= -\int_V \text{div}\,(V\mathbf{J})\mathrm{d}V + \int_V V\text{div}\,\mathbf{J}\mathrm{d}V.$$

The last term is zero and V is constant (steady state). Using Gauss's theorem the first term becomes

$$-V \oint_S \mathbf{n} \cdot \mathbf{J} \mathrm{d}A = -VI,$$

where I is the total current delivered by the source.

Equating the power absorbed in the resistor and the power delivered by the source for equal currents (charge conservation) we find:

$$I(u_1 - u_2) = -VI$$

and

$$V + (u_1 - u_2) = 0, \tag{3.59}$$

the sum of voltages must be zero. We recognize what is known as Kirchhoff's voltage law in circuit theory.

Now, we are going to consider a non-stationary current.

3.7 Time-Dependent Current Flow

Switching a current on also requires energy. A constant current causes a magnetic field; consequently a, in time changing, current will cause a, in time changing, magnetic field. This effect could not be taken into account in the case of a steady-state current.

The description of changing currents relies on a quasi-static fields description, that is, the fields only change a little during the time required for light to traverse a distance equal to the maximum dimension of the system under consideration. This is called the **quasi-stationary approach**.

The quasi-stationary approach takes into account a nonzero term $\partial \mathbf{H}/\partial t$, but $\partial \mathbf{E}/\partial t$ and $\partial \rho/\partial t$ are still considered to be zero, as is done in the steady-state approach. As a consequence, we can still use Kirchhoff's current law and a changing magnetic field can be taken into account to model time-dependent current flow in circuits.

Starting point is Faraday's law of induction equation 3.28,

$$\operatorname{curl} \mathbf{E} = -\mu_0 \frac{\partial \mathbf{H}}{\partial t}. \tag{3.60}$$

The vector potential is used to express the amount of energy necessary to build up the magnetic field. Starting with

$$W = \frac{1}{2}\mu_0 \int \mathbf{H}^2 \, dV \tag{3.61}$$

we find

$$W = \frac{1}{2} \int \mathbf{H} \cdot \text{curl} \, \mathbf{A} \, dV. \tag{3.62}$$

After partial integration we obtain

$$W = \frac{1}{2} \int \mathbf{A} \cdot \text{curl} \, \mathbf{H} \, dV = \frac{1}{2} \int \mathbf{A} \cdot \mathbf{J} \, dV. \tag{3.63}$$

Using equation 3.47 we find

$$W = \frac{\mu_0}{8\pi} \int \frac{\mathbf{J}_p \mathbf{J}_{p'}}{r_{pp'}} dV_p dV_{p'}. \tag{3.64}$$

This equation can be written as

$$W = \frac{1}{2}LI^2, \tag{3.65}$$

with I the time varying current through the circuit that causes the magnetic field. The quantity L, the **coefficient of self-inductance** is defined such that:

$$LI^2 = \frac{\mu_0}{4\pi} \int \frac{\mathbf{J}_p \mathbf{J}_{p'}}{r_{pp'}} dV_p dV_{p'}, \tag{3.66}$$

L only depends upon the geometry of the circuit.

Again consider a circuit with a conductor with resistance R and satisfying Ohm's law $\mathbf{J} = \sigma \mathbf{E}$. From Faraday's law and the definition of the vector potential it follows that

$$\text{curl}\,(\mathbf{E} + \frac{\partial \mathbf{A}}{\partial t}) = 0.$$

Defining a potential V leads to a generalization of the electrostatic relation

$$\mathbf{E} = -\frac{\partial \mathbf{A}}{\partial t} - \text{grad}\,V. \tag{3.67}$$

The heat produced in the conductor is again $I^2 R$. The source delivering

$$\int_V \mathbf{E} \cdot \mathbf{J} dV = -\int_V \text{grad}\,V \cdot \mathbf{J} dV - \int_V \mathbf{J} \cdot \frac{\partial \mathbf{A}}{\partial t} dV.$$

Using the same kind of manipulations as in the description of the energy conservation in case of a constant current and filling in equation 3.47 for the vector potential, one obtains:

$$\int_V \mathbf{E} \cdot \mathbf{J} dV = -V \oint_S \mathbf{n} \cdot \mathbf{J} dA - LI\frac{dI}{dt},$$

or

$$I^2 R = -IV - LI\frac{dI}{dt}$$

Thus, energy conservation leads to the important differential equation:

$$L\frac{dI}{dt} + RI = -V \tag{3.68}$$

where I is the total current delivered by the source. We see that for time-dependent currents the effect of a magnetic field is modelled by a self inductance. Again the sum of the voltages is zero if we associate V with the voltage of an energy source.

In case of a circuit with a capacitor an additional term is required. In circuits in which a conductor is used to charge a capacitor, the current is time-dependent and we can use equation 3.68 with a term relating the voltage across the capacitor, as a consequence of the electric field that is build up by the charge loaded on the capacitor, and the amount of charge Q stored on the capacitor with capacitance C. We obtain

$$L\frac{dI}{dt} + RI + \frac{Q}{C} = -V, \tag{3.69}$$

or, differentiated,

$$L\frac{d^2I}{dt^2} + R\frac{dI}{dt} + \frac{I}{C} = -\frac{dV}{dt}. \tag{3.70}$$

3.8 Towards Circuit Theory

In circuit theory, physical elements are represented by "lumped elements" without any extension. Real conductors are modelled by resistors, again without any extension. For time-dependent currents, the effect of changing magnetic fields can only be taken into account by adding the element inductor and in the same way changing electric fields can be taken into account by adding the element capacitor. Based on energy conservation the Kirchhoff's voltage law can be "derived" if any form of radiation is excluded.

This is a reasonable assumption because both the steady-state approach and the quasi-stationary approach at least require that $\partial \mathbf{E}/\partial t = 0$. In circuit theory, the voltages and currents are defined to follow the Kirchhoff current and voltage laws. These laws enter the theory as axioms.

Circuit theory and linear circuit theory, when only linear elements are considered, is a specialized discipline. Although loosely based on the quasi-stationary field approximation, it can often model real circuits quite accurately. However, it can also quite easily model circuits that do not represent any real physical circuit at all. The quality of results by modeling reality with (linear)circuits depends completely on the quality of the chosen equivalent circuits. Circuits with constant currents can be modelled with pure resistive circuits; in addition, time-dependent currents require elements that can store energy—such as (linear) inductors or (linear) capacitors. Some basics of circuit theory will be the topic of chapter 6.

Problems and Exercises

3.1 When an amount of charge is flowing out of a volume V, derive from the conservation of charge the continuity equation.

3.2 Show that curl grad $V = 0$, and that div curl $\mathbf{A} = 0$.

3.3 Prove $\operatorname{div}(\mathbf{E} \times \mathbf{H}) = -\mathbf{E} \cdot \operatorname{curl} \mathbf{H} + \mathbf{H} \cdot \operatorname{curl} \mathbf{E}$.

3.4 Discuss the Poynting vector applied to a capacitor that is being charged. (This is discussed for example in [Feynman *et al.* (1964)].)

3.5 What is the difference between the quasi stationary and steady-state approach?

Chapter 4

Free Electrons in Quantum Mechanics

In quantum mechanics electrons are described both as particles and as (matter) waves. Typical quantum mechanical phenomena such as energy quantization and tunneling are possible due to the wave nature of the electron. In the next two chapters an introduction to quantum mechanics is presented tailored for our needs. Especially, in this chapter the focus is on the quantum mechanical description of free electrons.

4.1 Particles, Fields, Wave Packets, and Uncertainty Relations

At the beginning of the last century electron scattering experiments of Davisson and Germer showed that a beam of electrons could produce interference patterns. This experiment, amongst many others, illustrates that electrons should have wave-like properties under certain circumstances.

According to quantum mechanics we associate a field—a matter field—with a single electron. This **matter field** or **wave function** $\Psi(x, t)$ describes the dynamic condition of an electron in the same sense that the electromagnetic field corresponds to photons which have precise momentum and energy. A **wavelength** λ and **frequency** f of the monochromatic field associated with an electron of momentum $\mathbf{p}$ and energy E are given by

$$\lambda = \frac{h}{p}, \qquad f = \frac{E}{h}, \tag{4.1}$$

where h is Plack's constant. These relations were first proposed by Louis de Broglie in 1924. Introducing the **wavenumber** $k = 2\pi/\lambda$ and the **angular frequency** $\omega = 2\pi f$, we may write the relations in the more symmetric

Table 4.1   Important quantities and symbols used in this chapter

Quantity	Symbol	Unit	Symbol
wavelength	λ	meter	m
frequency	f	per second	s^{-1}
energy	E	joule	J
Fermi energy	E_F	joule	J
wavenumber	k	per meter	m^{-1}
angular frequency	ω	per second	s^{-1}
momentum	$\mathbf{p}$	kilogram meter per second	kg·m/s
length	x	meter	m
time	t	second	s
mass	m	kilogram	kg
force	$\mathbf{F}$	newton	N
potential energy	E_P	joule	J
velocity	v	meter per second	m/s
current density	$\mathbf{J}$	ampère per square metere	A/m^2
Planck's constant	h	joule seconds	Js
Planck's constant divided by π	$\hbar$	joule seconds	Js
absolute temperature	T	kelvin	K
total number	N		
number per unit volume at a certain energy	N	per volume per energy	$m^{-3}J^{-1}$
number per unit volume	n	per volume	m^{-3}
fundamental charge	e	coulomb	C
Boltzmann constant	k	joule per kelvin	JK^{-1}
reflection coefficient	$\mathbf{R}$	ratio	
transmission coefficient	$\mathbf{T}$	ratio	

form

$$\frac{\mathbf{k}}{2\pi} = \frac{\mathbf{p}}{h}, \qquad \frac{\omega}{2\pi} = \frac{E}{h},$$

or, defining a new constant designated by $\hbar = h/2\pi$ we have

$$\mathbf{k} = \frac{\mathbf{p}}{\hbar}, \qquad \omega = \frac{E}{\hbar}. \tag{4.2}$$

Using the equations 4.1 and 4.2, we may represent the field corresponding to a free electron moving with a well-defined momentum $\mathbf{p}$ and energy E by a **harmonic wave** of constant amplitude.

Such a harmonic wave does not give any information about the localization in space of a free electron of well-defined momentum. An electron that is localized within a certain region Δx of space, however, should correspond to a matter field whose amplitude or intensity is large in that region

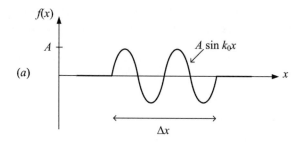

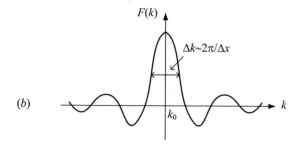

Fig. 4.1   (a) Harmonic pulse in x-space; (b) Fourier transform of the pulse in k-space

and very small outside it. The field of such a localized electron might correspond to a harmonic *pulse* in space, see also figure Fig. 4.1(*a*). From Fourier analysis we know that such a pulse can be obtained by superposing waves of different wavelengths. For this, quantum mechanics proposes to describe the localized electron as a **wave packet**.

The Fourier transform of the harmonic pulse is given by:

$$F(k) = \frac{1}{2}A\Delta x \left[\frac{\sin(\frac{1}{2}\Delta x \Delta k)}{(\frac{1}{2}\Delta x \Delta k)} \right]$$

Clearly, if the pulse had extended over the whole space, the power spectrum would only have one wavenumber k_0. In order to that the curve will zero outside the region Δx other wavenumbers have to be added so that the resultant Fourier series becomes zero. As a result, if the wave packet extends over a region Δx, the values of the wavenumbers of the interfering waves which compose the wave packet and have an appreciable amplitude, fall within the range Δk such that, see figure Fig. 4.1(b):

$$\Delta x \Delta k \sim 2\pi.$$

But, according to equations 4.1, different wavelengths λ mean that there is a range of allowed values of p such that $\Delta p = \hbar \Delta k$. The above expression becomes

$$\Delta x \Delta p \sim h. \tag{4.3}$$

The meaning of this relation is that the more precise our knowledge of the position of an electron, the more imprecise is our knowledge about its momentum, and conversely. Thus, an electron of well-known momentum is represented by a wave of constant amplitude extending over all space, and the knowledge of the position is nil. Relation 4.3 gives an optimum relation among the uncertainties Δx and Δp in the position x and momentum p of the electron. In most cases x and p are known with less accuracy, and relation 4.3 is written as

$$\Delta x \Delta p \geq \frac{\hbar}{2}, \tag{4.4}$$

known as one of **Heisenberg's uncertainty relations.**

In addition to this uncertainty relation between space and momentum, there is also an uncertainty relation between time and energy

$$\Delta t \Delta E \geq \frac{\hbar}{2}. \tag{4.5}$$

This relation, however, should *not* be interpreted in the meaning that a precise energy measurement requires a correspondingly long time. Energies can be measured fast with almost infinite precision. Still, it can be used, for example, to express the uncertainty in the energy of a particle in an excited level with a specified lifetime. This discussion shows that the interpretation of time is not that simple in quantum mechanics. This is also reflected in the existence of various definitions for the tunnel time of a tunneling electron.

4.2 Schrödinger's Equation

If m is the mass of the electron and j is the imaginary number: $\mathrm{j}^2 = -1$, in quantum mechanics the dynamics of the wave function $\Psi(x,t)$ of the electron is described by the Schrödinger equation. In one spatial dimension:

$$-\frac{\hbar^2}{2m}\frac{\partial^2}{\partial x^2}\Psi(x,t) + U(x)\Psi(x,t) = \mathrm{j}\hbar\frac{\partial}{\partial t}\Psi(x,t), \tag{4.6}$$

which is called the **time-dependent Schrödinger equation**. This equation describes an electron moving in a region of varying potential energy $U(x)$; as you see, forces do not enter directly.

Useful simplification is obtained by looking for separable solutions where the dependence on x and t is uncoupled. Therefore we write $\Psi(x,t) = \psi(x)\phi(t)$. Inserting this in the equation, we obtain

$$-\frac{\hbar^2}{2m}\frac{1}{\psi(x)}\frac{\partial^2\psi(x)}{\partial x^2} + U(x) = j\hbar\frac{1}{\phi(t)}\frac{\partial\phi(t)}{\partial t}. \tag{4.7}$$

The left-hand side only depends on position, the right only on time. Position and time are both independent variables, therefore the equation can only hold for all values of x and t only if both sides happen to be the same constant E. Equating the right-hand side to E we obtain the equation

$$\frac{d\phi(t)}{dt} = -\frac{jE}{\hbar}\phi(t), \tag{4.8}$$

which has the solution

$$\phi(t) = Ae^{-j(\frac{E}{\hbar})t} \tag{4.9}$$

In this oscillatory function the constant E has the dimension of energy and represents a well-defined energy $E = \hbar\omega$.

The total wave function is now given by

$$\Psi(x,t) = \psi(x)e^{-j(\frac{E}{\hbar})t}. \tag{4.10}$$

We see that quantum mechanical waves always have a complex time dependence.

4.2.1   *Time-independent Schrödinger equation*

Now, consider an electron. For the spatial part of $\Psi(x,t)$ we obtain a so-called **time-independent Schrödinger equation**. If we call $E_P(x)$ the one-dimensional potential energy of an electron and E is the total energy of the electron then (for one-dimensional problems)

$$-\frac{\hbar^2}{2m}\frac{d^2\psi(x)}{dx^2} + E_P(x)\psi(x) = E\psi(x), \tag{4.11}$$

where m is the mass of the electron.

The interpretation of $\psi(x)$, commonly also called **wave function**, is indirect: *the probability of finding the electron described by the wave function $\psi(x)$ in the interval dx around the point x is $|\psi(x)|^2 dx$.*

Because the electron must always be somewhere in space, the expression

$$\int_{all\ space} |\psi(x, y, z)|^2 \mathrm{d}x\mathrm{d}y\mathrm{d}z = 1 \qquad (4.12)$$

must hold. The expression is called the **normalization condition**. It implies severe limitations on the possible forms of the wave function. In particular, $\psi(x, y, z)$ must decrease very rapidly when the coordinates x, y, z become large, in order for the integral over all space to exist.

The wave function $\psi(x)$ is mostly expressed by a complex function. The complex conjugate of a complex function is obtained by replacing each j by (-j). Designating the complex conjugate of a function $\psi(x)$ by $\psi^*(x)$, then $|\psi(x)|^2 \equiv \psi^*(x)\psi(x)$ and is always real. Besides this, for any real $\psi(x)$ holds that $\psi(x) = \psi^*(x)$.

4.3　Free Electrons

In this section we investigate the description of free electrons in order to understand the behavior of conduction electrons in a metal (which are almost free) and tunneling electrons later on. In quantum mechanics electrons can be described as particles, as waves, or as wave packets. We inspect the various descriptions.

4.3.1　*Electron as a particle*

Before we consider the quantum mechanical wave function for a free electron we briefly look at the electron as a particle in classical mechanics. The dynamic state of an electron is determined by the forces acting on it and by the electron's total energy. The electron's momentum **p** is related to the force **F** on it by Newton's equation (we assume the velocity of the electron to be sufficiently far below the speed of light)

$$\mathbf{F} = \frac{\mathrm{d}\mathbf{p}}{\mathrm{d}t} \qquad (4.13)$$

and the electron's total energy is just the sum of its kinetic and potential energy

$$E = \frac{p^2}{2m} + E_P(x). \tag{4.14}$$

The velocity of the electron in the classical description becomes

$$\mathbf{v}_{cl} = \frac{\mathbf{p}}{m} \tag{4.15}$$

and can be used to calculate the mean velocity in the definition of the classical current density $\mathbf{J}$: if a charge distribution consist of electrons with charge e and moving with the mean velocity $\mathbf{v}$, then if N is the number of electrons per unit volume:

$$\mathbf{J} = N e \mathbf{v}. \tag{4.16}$$

4.3.2 *Electron as a wave*

In quantum mechanics the laws of conservation of momentum and energy remain valid. In case of a free electron the potential energy is zero and the one-dimensional Schrödinger's equation becomes

$$-\frac{\hbar^2}{2m}\frac{d^2\psi(x)}{dx^2} = E\psi(x). \tag{4.17}$$

For a free electron, $E = p^2/2m$. Setting $p = \hbar k$, we have

$$E = \frac{\hbar^2 k^2}{2m} \tag{4.18}$$

and equation 4.17 can be written as

$$\frac{d^2\psi(x)}{dx^2} + k^2\psi(x) = 0. \tag{4.19}$$

Direct substitution shows that the solutions

$$\psi(x) = e^{jkx} \quad \text{and} \quad \psi(x) = e^{-jkx}$$

are admitted. The wave function $\psi(x) = e^{jkx}$, now, represents a free electron of momentum $p = \hbar k$ and (kinetic) energy $E = p^2/2m$ moving in the $+X$-direction, and the wave function $\psi(x) = e^{-jkx}$ represents the free electron of the same momentum and energy in the opposite or $-X$-direction.

For either solution holds:

$$|\psi(x)|^2 = \psi^*(x)\psi(x) = e^{-jkx} \cdot e^{jkx} = 1.$$

This fact means that the probability of finding the electron is the same at any point, but describes a situation in which there is complete uncertainty about it's position.

The time-dependent wave function corresponding to $\psi(x) = e^{jkx}$ is

$$\Psi(x,t) = e^{j(kx-\omega t)}, \tag{4.20}$$

which describes an electron (as a wave) propagating in the $+X$-direction with a well-defined energy E and velocity $v = \omega/k = E/p$. In this case,

$$|\Psi(x,t)|^2 = [\psi^*(x)e^{jEt/\hbar}][\psi(x)e^{-jEt/\hbar}] = |\psi(x)|^2, \tag{4.21}$$

so that the probability density $|\Psi(x,t)|^2$ is independent of time. This is why states as these are designated as **stationary**.

Note, however, that these solutions are not the only solutions of the Schödinger equation, and that other solutions exist for which $|\Psi(x,t)|^2$ is not independent of time. The corresponding states are not stationary. The Schödinger equation is a linear homogeneous differential equation, and thus, the general solution corresponding to nonstationary states can be expressed by a linear combination of stationary-state solutions

$$\Psi(x,t) = \sum_n c_n \psi_n(x) e^{-jE_n t/\hbar}. \tag{4.22}$$

As an example consider $\Psi(x,t)$ as the sum of two stationary terms

$$\Psi(x,t) = c_1 \psi_1(x) e^{-jE_1 t/\hbar} + c_2 \psi_2(x) e^{-jE_2 t/\hbar}. \tag{4.23}$$

The probability distribution $|\Psi(x,t)|^2$ is, now, not constant but contains complex terms that oscillate with the angular frequency $\omega = (E_1 - E_2)/\hbar$ or with the linear frequency $\nu = (E_1 - E_2)/h$. The wave function may describe a system during the transitions of an electron between stationary states E_1 and E_2. We may say that the electron is oscillating with a frequency $\nu = (E_1 - E_2)/h$, and therefore capable of emitting or absorbing electromagnetic radiation of the same frequency. More detailed description is not necessary here, but can be found in introductory textbooks on quantum mechanics, for example in [Alonso and Finn (1967)].

4.3.3 *A beam of free electrons*

Until now we considered single electrons and described them with a wave function. Experiments show that a monochromatic wave of a single wavelength $\lambda = h/p$ can be associated with a beam of electrons too. If a single electron is described by a monochromatic wave, then it's position is unknown and the probability of finding the electron is the same at any point of space. So, obviously, if we have a beam of a large number of electrons the distribution of those electrons in the whole beam can assumed to be uniform.

We consider, therefore, the scalar ψ to be capable of representing in a statistical fashion the behavior of each electron in the beam. It will likewise contain a description of the statistical effects of the superposition of a large number of electrons of the same energy making up the electron beam. Let us consider a electron beam to be monoenergetic and of uniform density, and choose a coordinate system so that the electrons in the beam are traveling in the $+X$-direction. If this beam is to be described by a single plane wave then [Fromhold (1981)]

$$\Psi(x, t) = A e^{j(kx - \omega t)}, \qquad (4.24)$$

The complex form chosen for $\Psi(x, t)$ assures that the spatial density of electrons $|\Psi|^2 = |A|^2$ is uniform; this would not be the case if we had chosen the real function $\Psi(x, t) = A \cos(kx - \omega t)$, for example. Now, the intensity of the beam (that is, the number of electrons per unit volume) can be expressed by $|A|^2$. The flux (that is, the electron current density) is $v|A|^2$.

Substitution of $k = p/\hbar$ and $\omega = E/\hbar$ gives:

$$\Psi(x, t) = A e^{(j/\hbar)(px - Et)}, \qquad (4.25)$$

Suppose we differentiate Ψ with respect to t and x,

$$\frac{\partial \Psi}{\partial t} = -(\frac{j}{\hbar})E\Psi \qquad (4.26)$$

$$\frac{\partial \Psi}{\partial x} = (\frac{j}{\hbar})p\Psi \qquad (4.27)$$

$$\frac{\partial^2 \Psi}{\partial x^2} = -(\frac{p^2}{\hbar^2})\Psi \qquad (4.28)$$

Solving for p^2 and E and substituting into $p^2/2m = E$ gives

$$-\frac{\hbar^2}{2m}\frac{\partial^2 \Psi}{\partial x^2} = j\hbar\frac{\partial \Psi}{\partial t}. \qquad (4.29)$$

Thus a flux of free electrons satisfies the time-dependent Schrödinger equation for free electrons $(U(x) = 0)$.

Leaving out the time behavior, the general solution can be written as a linear combination of solutions in two directions, that is,

$$\psi(x) = Ae^{jkx} + Be^{-jkx}, \qquad (4.30)$$

with A and B arbitrary complex numbers, now representing beams of electrons going in both $+X$- and $(-X)$ directions, in which the electrons have the same energy.

4.3.4   *Electron as a wave packet*

As we saw in subsection 4.3.2 the position of an electron described by a single well-defined momentum is not known. This is in agreement with the uncertainty principle. As discussed in section 4.1, we can obtain information about the position of an electron localized within a region Δx when we superpose several solutions of the form Ae^{jkx}, with different values of k (or p), and with appreciable amplitude A in the range Δk (or Δp); that is, we must form a wave packet. This wave packet can be expressed as

$$\psi(x) = \int A(k)e^{jkx}dk, \qquad (4.31)$$

where $A(k)$ is the amplitude corresponding to the momentum $p = \hbar k$.

4.3.5   *Phase- and group velocity*

Using the equation for the energy of the free electron and its momentum when the electron is described using waves, equations 4.2 and 4.18, we find the dispersion relation between frequency and wave number to be $\omega = (\hbar/2m)k^2$. This is a nonlinear relation, which means that the velocity of the electron waves is a function of their frequency and must be defined carefully. We distinguish two velocities: the phase velocity

$$v_{\mathrm{ph}} = \frac{\omega}{k} = \frac{\hbar k}{2m} = \frac{p}{2m} \qquad (4.32)$$

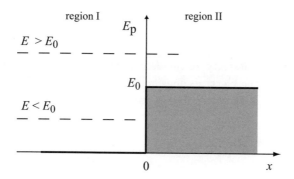

Fig. 4.2 Step potential with the total energy per electron of the incident beam denoted by E. The classical forbidden area is shaded.

and the group velocity

$$v_g = \frac{d\omega}{dk} = \frac{\hbar k}{m} = \frac{p}{m} = v_{cl}, \qquad (4.33)$$

where v_{cl} is the classical velocity of a particle. We are usually interested in the behavior of the wave packet as a whole rather than its internal motion. The group velocity is then the appropriate one and as we see this agrees with the classical result.

4.4 Free Electrons Meeting a Boundary

As a first illustration of the use of the Schrödinger equation, we determine the wave function $\psi(x)$ for an electron moving in a region in which the potential energy is illustrated as in figure Fig. 4.2; the so called **step potential**. The potential energy is zero for $x < 0$ (region I) and has a constant value E_0 for $x > 0$ (region II).

We choose an incident beam of electrons

$$\Psi_{inc}(x,t) = Ae^{j(kx-\omega t)} \qquad (4.34)$$

traveling to the right in region I towards the step at $x = 0$. The incident beam consists of individual electrons having each a well-defined momentum $p = \hbar k$ in the x-direction and a well-defined kinetic energy $E_K = p^2/(2m)$ equal to the total energy E, since the potential energy E_P is zero in region I. The electron density in the incident beam is given by

$$\Psi_{\text{inc}}^* \Psi_{\text{inc}} = A^* A = |A|^2. \tag{4.35}$$

The incident beam intensity, or incident electron "flux", I_{inc} will be given by the product of the electron density and the electron velocity $v = p/m$ (and gives the number of electrons passing through a unit area per unit time):

$$I_{\text{inc}} = \Psi_{\text{inc}}^* \Psi_{\text{inc}} v = A^* A \left(\frac{\hbar k}{m}\right), \tag{4.36}$$

where

$$\hbar k = p = \sqrt{2mE} \tag{4.37}$$

The positive sign is to be chosen for the square root. The total energy E of the electron is the same in region I as in region II because there are no external forces acting on the electron. The step potential does instantaneously accelerate the electron in the negative direction, which leads to an interchange of kinetic and potential energies for the conservative system. It is necessary to consider separately the cases for which $E < E_0$ and for which $E > E_0$

4.4.1 Step potential: $E < E_0$

This is a model that is suitable for free electrons reaching the boundaries in a metal. All electrons having an energy lower than the work function cannot escape from the metal; when they reach the surface, they are turned back into it. Consider electrons with $E < E_0$. In this case, classical mechanics tells us that the electrons cannot be in region II, because there their kinetic energy $E_K = E - E_P$ would be negative, which is impossible. Thus, $x > 0$ is a classically forbidden region. To obtain the total $\psi(x)$ for a step potential we must write down Schrödinger's equation separately for regions I and II. In region I, in which $E_P(x) = 0$ we obtain

$$\frac{d^2 \psi_1}{dx^2} + \frac{2mE}{\hbar^2} \psi_1 = 0 \tag{4.38}$$

This is the same equation as that for free electrons and the solution will be the same.

$$\psi_1(x) = A e^{jkx} + B e^{-jkx} \quad , \qquad k^2 = \frac{2mE}{\hbar^2} \tag{4.39}$$

Written in this way it represents an incident beam with electrons and a reflected beam with electrons. We expect electrons to reflect in this case. In classical physics we expect all electrons to reflect. In this quantum mechanical calculation, however, we will start to assign a different amplitude to the reflected beam to take into account any possible change of the incident beam as a result of the reflection at $x = 0$.

In region II, in which $E_P(x) = E_0$, the kinetic energy $E_K = E - E_0$ would be negative. We can, however, still obtain a solution of the Schrödinger equation

$$\frac{d^2\psi_2}{dx^2} + \frac{2m(E - E_0)}{\hbar^2}\psi_2 = 0. \tag{4.40}$$

If we define a positive quantity $\alpha^2 = 2m(E_0 - E)/\hbar$ equation 4.40 becomes

$$\frac{d^2\psi_2}{dx^2} - \alpha^2\psi_2 = 0 \quad , \quad \alpha^2 = \frac{2m(E_0 - E)}{\hbar^2}. \tag{4.41}$$

The solution of this differential equation is a combination of the functions $e^{\alpha x}$ and $e^{-\alpha x}$, as you may verify by direct substitution. The increasing function $e^{\alpha x}$ is not acceptable because it diverges for $x \to \infty$, and we know that the wave amplitude must be very small in region II; experience tells us that we are not likely to find an electron in this region. We obtain as a useful solution

$$\psi_2(x) = Ce^{-\alpha x}. \tag{4.42}$$

Note also that $\psi_2(x) = e^{-\alpha x}$ cannot be interpreted as representing electrons moving to the left. Real exponentials do not represent waves with a wavelength, and with no wavelength there is no real momentum.

We can determine the constants A, B, and C only by applying the condition of continuity of the wave function at $x = 0$, which is an obvious physical argument. Also the wave function must change smoothly as it crosses the potential step. This requires that

$$\psi_1 = \psi_2 \quad \text{and} \quad \frac{d\psi_1}{dx} = \frac{d\psi_2}{dx} \quad \text{for} \quad x = 0. \tag{4.43}$$

These conditions yield

$$A + B = C \tag{4.44}$$

$$jk(A - B) = -\alpha C, \tag{4.45}$$

or equivalently

$$A + B = C \tag{4.46}$$

$$A - B = -\frac{\alpha}{jk}C. \tag{4.47}$$

These two equations with three unkowns give in turn

$$B = \frac{jk + \alpha}{jk - \alpha}A \quad \text{and} \quad C = \frac{2jk}{jk - \alpha}A, \tag{4.48}$$

so that

$$\psi_1(x) = A\left(e^{jkx} + \frac{jk + \alpha}{jk - \alpha}e^{-jkx}\right) \quad \text{and} \tag{4.49}$$

$$\psi_2(x) = \frac{2jk}{jk - \alpha}Ae^{-\alpha x}. \tag{4.50}$$

The fact that $\psi_2(x)$ is not zero means that there is some probability of finding the electron in region II. That is, in quantum mechanics, the regions in which the electron may move does not, in general, have sharp boundaries. Since $\psi_2(x)$, however, is given by a decreasing exponential the probability of finding the electron in region II decreases very rapidly as x increases. In general, therefore, the electron cannot go very far into the classically forbidden region. In a later chapter we will see that currents (and thus beams) can exist only in region I and not in region II; that is all electrons reaching the potential step with $E < E_0$ bounce back, including those that penetrate slightly into region II. We can support this assertion by calculating the intensity of the reflected beam $\hbar(-k)/m|B|^2$.

The intensity of the incoming beam is $\hbar k/m|A|^2$. The intensity of the reflected beam is

$$v|B|^2 = \frac{\hbar(-k)}{m}|B|^2 = \frac{\hbar(-k)}{m}\left|\frac{jk + \alpha}{jk - \alpha}A\right|^2 = \tag{4.51}$$

$$= \frac{jk + \alpha}{jk - \alpha} \cdot \frac{-jk + \alpha}{-jk - \alpha}\frac{\hbar(-k)}{m}|A|^2 = (-v)|A|^2,$$

the incident and reflected electrons move with the same, but opposite, velocity. We see that both the incident and the reflected beams have the same intensity in absolute values.

4.4.2 Step potential: $E > E_0$

In the case that $E > E_0$, if we assume that all the electrons are coming from the left, the classical description would be that all electrons proceed into region II, although they move with a smaller velocity v' than in region I; the momentum of an electron in region II is $p' = [2m(E - E_0)]^{1/2}$. At $t = 0$ the electron suffers a sudden deceleration. The solution for region I is still given by equation 4.39, $\psi_1(x) = Ae^{jkx} + Be^{-jkx}$, if we assume that it is possible that some electrons are reflected (an assumption we will verify). For region II the solution differs from the previous case, because electrons can pass the barrier and will thus form a transmitted beam. We expect the wave function for this transmitted beam to be of the form

$$\Psi_{\text{trans}}(x,t) = Ce^{j(k'x-\omega t)} \quad , \quad (k')^2 = \frac{2m(E - E_0)}{\hbar^2} \tag{4.52}$$

The alternate form of a reverse-traveling wave in region II might also be expected to satisfy the Schrödinger equation, but this wave will not be considered here because in the present problem there is no source of electrons in region II and there are no additional barriers in region II to give reflection of the forward-propagating beam. Thus for the present problem,

$$\psi_2 = \psi_{\text{trans}} \tag{4.53}$$

The application of the two boundary conditions provides again two equations to determine the constants B and C for the reflected and transmitted wave amplitudes in terms of the incident wave amplitude A.

$$A + B = C \tag{4.54}$$

$$jk(A - B) = jk'C, \tag{4.55}$$

or equivalently

$$A + B = C \tag{4.56}$$

$$A - B = \frac{k'}{k}C. \tag{4.57}$$

The solutions are

$$B = \frac{(k - k')}{k + k'}A \quad \text{and} \quad C = \frac{2k}{k + k'}A, \tag{4.58}$$

so that

$$\psi_1(x) = A\left(e^{jkx} + \frac{k - k'}{k + k'}e^{-jkx}\right) \quad \text{and} \tag{4.59}$$

$$\psi_2(x) = \frac{2k}{k + k'}Ae^{jk'x}. \tag{4.60}$$

The fact that B is not zero proofs that some electrons are reflected at $x = 0$. We can, now, calculate the so-called **reflection coefficient R** and **transmission coefficient T**, that is, the absolute value of the quotient of the intensity of the reflected beam and the intensity of the incident beam, and the absolute value of the quotient of the intensity of the transmitted beam and the intensity of the incident beam, respectively.

$$\mathbf{R} = \left| \frac{(-v)|B|^2}{v|A|^2} \right| = \left(\frac{k - k'}{k + k'} \right)^2 \tag{4.61}$$

$$\mathbf{T} = \left| \frac{v'|C|^2}{v|A|^2} \right| = \frac{k'}{k} \left(\frac{2k}{k + k'} \right)^2 = \frac{4kk'}{(k + k')^2}. \tag{4.62}$$

We see that **R** and **T** are smaller than 1, because the incoming beam of electrons is split into the transmitted and reflected beams. One can verify easily that $\mathbf{R} + \mathbf{T} = 1$, which is required for conservation of the number of electrons (conservation of charge), since the number of incoming electrons must be equal to the sum of those reflected and those transmitted.

4.5   Electrons in Potential Wells

In other applications the motion of an electron is restricted by an external force or interaction to a finite region in space. An important example is an electron in a metal or on a small island. One can say the electron is bounded in space in these cases, and as we will see, this gives rise to bounded states. Consider the situation in which the electron is confined to a small finite region of space. This is a model for what is also called a **quantum well**, **quantum dot**, or just particle in a potential box.

4.5.1   *Infinite well: standing waves*

The simplest example is a infinitely deep square well: the electron has zero potential energy in the region $0 < x < a$, and infinitely high barriers that prevent it from leaving this region. This *is* a model of a free electron in a wire: a conduction electron can move freely through a metal, but it cannot escape from it. It is an oversimplified model, of course, since the energy the electrons need to escape from the metal is not infinite but equals the electron charge times the metal's work function. However, as we will see the model describes the metallic structures quite well.

The Schrödinger equation inside the one-dimensional well is identical to that for the free electron, because inside the well the potential energy is zero and the electron only possesses kinetic energy. The solutions are therefore the same, and can be written in terms $\exp(+jkx)$ and $\exp(-jkx)$, with $E = \hbar^2 k^2/2m$, however the correct solution should met the boundary conditions.

Between $x = 0$ and $x = a$ the wave function is

$$\psi(x) = Ae^{jkx} + Be^{-jkx},$$

expressing motion in both directions. The barriers present an infinitely high potential energy outside the well, and consequently outside the well $\psi(x)$ must be zero to force the term $V(x)\psi(x)$ to remain finite. Because discontinuous wave functions are unacceptable $\psi(x)$ must vanish at the edge of the barriers. This implies boundary conditions $\psi(x = 0) = 0$ and $\psi(x = a) = 0$. Then

$$A + B = 0 \quad \text{or} \quad B = -A$$

So

$$\psi(x) = A(e^{jkx} - e^{-jkx}) = 2jA \sin kx = C \sin kx, \tag{4.63}$$

where $C = 2jA$ and might be complex. Knowing that C cannot be zero we find that $\sin kx = 0$ at the boundaries or

$$k = \frac{n\pi}{a} \quad \text{or} \quad p = \hbar k = \frac{n\pi\hbar}{a}, \tag{4.64}$$

p is the momentum of the electron. The electron's energy is now given by

$$E = \frac{p^2}{2m} = \frac{\hbar^2 k^2}{2m} = \frac{n^2\pi^2\hbar^2}{2ma^2}. \tag{4.65}$$

The conclusion is that the electron cannot have any arbitrary energy, but only those values given by equation 4.65, as shown in figure Fig. 4.3. As we see, the energy is quantized. The quantization of energy comes from the boundary conditions that constrain the motion of the electron inside the well. The lowest state has a nonzero energy, and is called the **zero-point energy**.

The wave functions corresponding to the k values are

$$\psi_n = C \sin(\frac{n\pi x}{a}) \tag{4.66}$$

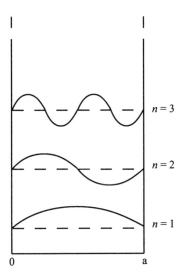

Fig. 4.3   Solutions of the Schrödinger equation for an electron in a 1-dimensional infinite quantum well

and the integer $n = 1, 2, 3, \ldots$ is a **quantum number** that labels the states.

Going back to modelling of a metallic structure we notice that electrons that move freely over all space have a continuous range of energy levels; those that are confined to a finite region of space, such as electrons in a metallic wire, have discrete levels. When, however, the width of the region in space becomes very large the separation of the energy levels will become very small and approximation by a continuous range again may be possible. We also see that the wave functions $\psi_n(x)$ are expressed by a (complex) number times a real function in this case, and as we will see in the next chapter, such a wave function alone cannot carry a current [Davis (1998)]. Of course we do know that a metal conducts current, when a voltage is applied over its terminals. The solution is that a superposition of wave functions is needed to conduct a current in a wire, meaning that the conduction electrons will oscillate between (empty) energy states, as was discussed in subsection 4.3.2. The necessary energy is provided by the electromagnetic field produced by an energy source (or in other words we need a non-zero voltage across a wire to conduct a current).

4.5.2 *Finite well: periodic boundary conditions*

A more suitable model for a metallic structure seems to be a finite well, because electrons can escape from a wire when their energy is large enough. The finite square well will be discussed briefly.

In the well the solution of the Schrödinger equation is identical to the corresponding result in the infinite case, the electrons are still considered to be free. Outside the well, however, the Schrödinger equation becomes

$$\frac{\hbar^2}{2m}\frac{d^2\psi}{dx^2} + (E_0 - E)\psi = 0 \qquad (4.67)$$

whose general solution is

$$\psi = C\exp(\kappa x) + D\exp(-\kappa x) \qquad (4.68)$$

where C and D are (complex) constants and $\kappa = [2m(E_0 - E)/\hbar^2]^{1/2}$. C or D must equal zero to prevent the wave function to become infinite as x goes to infinity and guaranties that there will be no current outside the well. Thus we have outside the well

$$\psi = D\exp(-\kappa x). \qquad (4.69)$$

if x is positive and

$$\psi = D\exp(\kappa x). \qquad (4.70)$$

if x is negative. As the discontinuities at the boundaries of the well are now finite rather than infinite, the boundary conditions require ψ and its derivative to be continuous at these points. The expressions for the wave functions can, however, not be obtained analytically. Usually a graphical method of finding possible solutions is used and can be found in standard textbooks on quantum mechanics. Only the results of such a procedure will be presented here.

Just as in the case of the infinite well the wave function is a single function and currents in the well can only be described by superposition of states. Comparing the wave functions in the well with the wave functions for the infinite well, they are generally similar. An important difference is that in case of the finite well, the wave functions decay exponentially outside the well, implying that there *is* a probability of finding the electron just outside the well. The wave function penetrates into the classical forbidden region and results in the energy levels being lower than in the infinite well case. For energy levels far below E_0 this last effect is hardly noticeable.

4.5.3   *Quantization of energy*

We saw that energy was quantized in both cases of the potential well. This situation is not a peculiarity of the potential well, it generally holds whenever Schrödinger's equation is solved for a potential energy that confines the electron to move in a limited region. Energy quantization is due to the fact that the wave function is determined by both the potential energy and the boundary conditions. An acceptable wave function in these cases in general exists only for certain values E_1, E_2,... of the energy. It is the mathematical formulation of quantum mechanics with Schrödingers equation that incorporates the quantization of energy and the existence of a discrete set of allowed quantum states.

We may ask how many quantum states exist in a small energy range in a very large potential well. Therefore we consider a cubical potential box of side a. The energy levels in this 3-dimensional case are given by generalizing equation 4.65

$$E = \frac{\pi^2 \hbar^2}{2ma^2}(n_1^2 + n_2^2 + n_3^2) \qquad (4.71)$$

where n_1, n_2, and n_3 are integers (n_1, n_2, and n_3 are quantum numbers). For a small box the energy levels are spaced widely, but for a large box, as in the case of electrons in a metal, successive levels are so close that they practically form a continuous spectrum. We also see that for different combinations of the quantum numbers the same energy level is calculated. However, each set of quantum numbers (or quantum state) corresponds to a different wave function. The coincidence that different wave functions have the same energy is called **degeneracy**.

Following the general accepted symbols for the various quantities, accepting that N will be used for both the total number of states and the number of states per unit volume, we can find an expression for how many states there are in a small energy range dE, in the almost continuous spectrum case. Therefore, a new number κ is defined: $\kappa^2 = n_1^2 + n_2^2 + n_3^2$. It can be used to tell us how many quantum states can be found in a sphere of radius κ in the 3-dimensional space with coordinates n_1, n_2, and n_3

$$E = \frac{\pi^2 \hbar^2}{2ma^2}\kappa^2. \qquad (4.72)$$

To find the number of quantum states $N(E)$ with energy between the zero-point energy and E, we must find the volume of an octant of the sphere of radius κ, since n_1, n_2, and n_3 can only have positive values. We obtain

$$N(E) = \frac{1}{8}(\frac{4}{3}\pi\kappa^3) = \frac{\pi}{6}V\left(\frac{2mE}{\pi^2\hbar^2}\right)^{3/2}, \tag{4.73}$$

where $V = a^3$ is the volume of the potential box. Using $\hbar = h/2\pi$ we get

$$N(E) = \frac{8\pi V}{3h^3}\left(2m^3\right)^{1/2}E^{3/2}. \tag{4.74}$$

The number of states with energy between E and $E + dE$ is obtained by differentiating the above equation, yielding

$$dN(E) = \frac{4\pi V\left(2m^3\right)^{1/2}}{h^3}E^{1/2}dE.$$

It is convenient to define the number of states per unit energy interval at the energy E, the **density of states per spin direction** $\tilde{g}(E)$, $\tilde{g}(E) = dN/dE$, so that

$$\tilde{g}(E) = \frac{dN}{dE} = \frac{4\pi V\left(2m^3\right)^{1/2}}{h^3}E^{1/2}. \tag{4.75}$$

4.5.4 *Free-electron model*

We can use the density of states, equation 4.75, to find the number of free electrons per unit volume in a solid at a certain energy N. Knowing that two electrons can occupy each energy level (one with spin up and one with spin down) we have:

$$N = \frac{2\tilde{g}(E)}{V} = 2\frac{dN}{dE} = \frac{8\pi\left(2m^3\right)^{1/2}}{h^3}E^{1/2}. \tag{4.76}$$

And, the number of electrons per unit volume that can be accommodated up to the energy E, n, is given by equation 4.74 multiplied by two:

$$n = \frac{2N(E)}{V} = \frac{16\pi}{3h^3}\left(2m^3\right)^{1/2}E^{3/2}. \tag{4.77}$$

At zero temperature, the electrons fill all energy levels upto the last level called Fermi level and the three dimensional sphere is called the Fermi sphere. An expression for the energy at the Fermi-level can be found using equation 4.73 and realizing that there are two electrons in each quantum state at $T = 0$ K:

$$E_F = \frac{h^2}{8m} \left(\frac{3N}{\pi V} \right)^{2/3} \tag{4.78}$$

We can also relate the number of electrons per unit volume upto the Fermi level, n, to the number of electrons per unit volume at the Fermi level, $N(E_F)$, using equations 4.76 and 4.77:

$$n = \frac{2}{3} E_F N(E_F), \tag{4.79}$$

this relation will be used later.

In metals the conduction electrons are relatively free to move through the solid. Because their mutual repulsion is cancelled, on the average, by the attraction of the lattice atoms, we may regard the conduction electrons as approximately free electrons and can treat them to good approximation as a classical gas, a "free-electron gas".

In order to calculate the number of free electrons, at nonzero temperatures, with respect to energy in the potential well, we have to discuss briefly the statistics governing the distribution of energy among thermally agitated electrons. We assume that the thermally agitated free electrons behave like in a gas. This approximation is usually called the free-electron model of a solid.

The basic description comes from thermodynamics and tells us that the density of gas particles decreases exponentially with the energy. The energy is measured in kT energy units; k is Boltzmann's constant and T the absolute temperature. A **relative probability** can be defined

$$f_B(E) = \frac{N}{N(E_0)} = \exp^{-(E-E_0)/kT} \tag{4.80}$$

Where E_0 is some energy reference. $f_B(E)$ is called the Boltzmann distribution and is valid for the energy distribution of gasses in general regardless whether the energy is potential or kinetic.

The important question is whether it applies to a free electron gas too. As we will see the free electrons do follow the Boltzmann distribution at high energies. At low energies things are different. A limitation is imposed by the Pauli exclusion principle that does not exist in a classical gas.

As we saw in this section, an infinite potential well has an almost continuum of discrete energy levels from the bottom of the well upwards. At zero temperature energy levels upto the Fermi level will be filled (two electrons on each level as mandated by the Pauli exclusion principle); above these levels all others are empty. At increasing temperatures we expect that the lower levels are still almost filled, while an increasing number of electrons are in levels above the Fermi level. We expect that at high-energy levels the electron distribution follows the Boltzmann distribution. This is because at high energies the restriction to one electron per state has no significance, since there are far more states than electrons.

A statistical treatment of the electron distribution based on the one-electron-per-state condition results in the **Fermi-Dirac distribution**. This fairly involved treatment will not be done here; the well-known result is

$$f(E) = \frac{1}{\exp^{(E-E_\mathrm{F})/kT} + 1}. \tag{4.81}$$

At low temperatures, i.e. $T \to 0$, for energies smaller than E_F the exponential term becomes zero and $f(E)$ is unity up to E_F; all energy levels upto the Fermi level are filled completely. The Fermi-Dirac distribution has the property that $f(E) = 1/2$ at the Fermi energy. And finally, it coincides with the Boltzmann distribution at high energies.

The knowledge of the distribution of electrons over the energy levels in a potential well allows us to find the number of electron in a metal in a certain energy band. The number of electrons in such a band is determined by twice the sum of the available states times the probability that the state is filled. This is how equation 2.6 in chapter 2 was obtained.

4.5.5 Quantum cellular automata (QCAs)

Based on the above described quantum wells nanoelectronic devices have been proposed and manufactured. One example is the group of quantum cellular automata. Quantum cellular automata (QCA) are nanoelectronic digital logic architectures in which information is stored and processed as configurations of pairs of electrons in arrays of . Quantum cellular automata have been originally proposed by Lent *et.al.* [Lent *et al.* (1993)]. The basic QCA cell consists of four small quantum wells (also called quantum dots) in a square separated by tunnel barriers. Each cell contains two additional free electrons. Coulomb interaction forces these electrons to dots on opposite corners. This results in two different configurations and are corresponding to binary "0" and "1", see figure Fig. 4.4. Changing the inputs will result in a new output. The example shows the majority of three inputs. In this

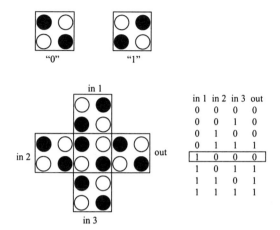

Fig. 4.4 Quantum cellular automata consists of arrays of coupled quantum wells. Presence and absence of additional electrons in quantum dots cause "logic" functions to propagate through the array. As an example a logic threshold function is shown.

case only one input is "1" while the others are both "0". Due to Coulomb repulsion of the additional electrons, the distribution of the electrons in this case will force the output to be "0", the "majority" of the inputs. When the input is changed to have two "1"s the output will be changed to "1" (check this for yourself!). An obvious disadvantage of logic functions made with quantum cellular arrays is that they are restricted to specific topologies. More complex functions (for example, a full adder) will therefore use much space. The structures have the same problems as the structures that were used in the "old days" (the 1970-80's) of charge-coupled device (CCD) logic [Zimmerman *et al.* (1977); van der Klauw (1986); Hoekstra (1987)].

Problems and Exercises

4.1 The position of an electron is determined with an uncertainty of 0.1 nm. (a) Find the uncertainty in its momentum. (b) If the electron's energy is in the order of 10 eV, estimate the uncertainty in its energy.

4.2 What is the effect on the energy levels of a one-dimensional potential box as the size of the box (a) decreases? (b) increases?

4.3 Show that **R** + **T** = 1 for a beam of electrons impinging on a potential step if the energy of the electron is larger than the potential step.

4.4 Show that the fractional energy difference $\Delta E/E$ between any two adjacent levels of a particle in a box is given by $(2n+1)/n^2$.

4.5 Design using a QCA (1) an AND function and (2) an OR function.

Chapter 5

Current and Tunnel Current in Quantum Physics

In chapter 3 we looked at the description of current in classical electrodynamics and circuit theory. In this chapter the quantum physics approach
is discussed. Such a approach is necessary because for microscopic device
dimensions the classical theory does not correctly describe reality. The current will be discussed in case of both macroscopic and microscopic device
dimensions.

5.1 Electrical Conductivity in Metals

In metals the current is defined as the rate of flow of conduction electrons
through a cross-sectional area. Experimentally, we know that for metals
Ohm's law hold for not too high electric fields and that the resistivity
increases linearly with temperature. A microscopic picture of the motion
and the position of the conduction electrons is, however, not that simple.

In the classical theory of metallic conduction at room temperature free
electrons move in random directions with relatively large speeds of the order
of 10^5 m/s if there is no electric field in the wire; since the velocity vectors
are randomly oriented the average velocity is zero. When there is an electric
field in the wire as a result of the connection to an electric energy source,
a free electron (a conduction electron) feels a acceleration due to the force
$(-e)\mathbf{E}$, and acquires an additional velocity in the direction opposite the field.
Because the electrons are not endlessly accelerated, an scattering process
must be involved. The classical picture, presented by Drude in 1900 and
elaborated in more detail by Lorentz in 1905, is that the kinetic energy
acquired by the accelerated electron is quickly dissipated by collisions with
the lattice ions in the metal. The net result of the repeated acceleration
is that the electron has a small average velocity called **drift velocity** $\mathbf{v}_d$

opposite to the electric field.

In a metal, measurements on the **mean free path** Λ, that is: the average distance the electron travels between collisions or scattering events, and on the temperature dependence, however, are not in agreement with the predictions based on the classical model. More complicated models based on quantum mechanics are necessary to see that (1) we have to use the free-electron model of metals in which the conduction electrons are located only at the Fermi level, (2) scattering arises because of imperfections of the crystal lattice, and (3) scattering arises because of interaction with phonons—quantized thermal vibrations of the lattice ions. For our purpose the most important conclusion will be that in metals the conduction electrons move only in the direct neighborhood of the Fermi level. It is not necessary to go in full details of all derivations, but the main line of reasoning will be presented.

5.1.1 *Drude model*

To start the discussion on the basic properties, in the Drude model, the conduction electrons in the metal are treated as charged particles that move freely in the lattice and collide with lattice atoms. If an electric field E_x is applied along the x-direction the equation of motion for each electron is expressed as:

$$\frac{\mathrm{d}p_x}{\mathrm{d}t} = \frac{\hbar \, \mathrm{d}k_x}{\mathrm{d}t} = (\text{-e})E_x. \tag{5.1}$$

The solution of equation 5.1 is

$$k_x = [(\text{-e})E_x/\hbar]\, t + k_0 \tag{5.2}$$

and shows immediately that the wave number k_x increases indefinitely with increasing time. This means that the Fermi sphere moves as a whole in the direction opposite to the applied field and the electrical current becomes infinitely large. Of course this does not happen in a real metal. Instead, a steady current flows as long as a constant field is applied; we measure an ohmic behavior. Thus, a scattering process must be involved. However, the electrons are *not* scattered by the ions in the lattice. To understand this we first consider what would have happened if conduction electrons were scattered by ions in the lattice.

To introduce a scattering term in equation 5.1 a drift velocity $\mathbf{v}_d$ is defined as the velocity per electron averaged over a whole assembly of con-

Table 5.1 Important quantities and symbols used in this chapter

Quantity	Symbol	Unit	Symbol
wavelength	λ	meter	m
frequency	f	per second	s^{-1}
energy	E	joule	J
Fermi energy	E_F	joule	J
wavenumber	$\mathbf{k}$	per meter	m^{-1}
angular frequency	ω	per second	s^{-1}
momentum	$\mathbf{p}$	kilogram meter per second	kg·m/s
length	x	meter	m
mean free path	Λ	meter	m
time	t	second	s
mass	m	kilogram	kg
force	$\mathbf{F}$	newton	N
potential energy	E_P	joule	J
velocity	$\mathbf{v}$	meter per second	m/s
acceleration	$\mathbf{a}$	meter per square second	m/s^2
temperature	T	kelvin	K
number per unit volume at a certain energy	N	per volume per energy	$m^{-3}J^{-1}$
number per unit volume	n	per volume	m^{-3}
charge	q	coulomb	C
charge density	ρ	coulomb per cubic meter	C/m^3
electric field strength	$\mathbf{E}$	volt per meter	V/m
current	I	ampère	A
current density	$\mathbf{J}$	ampère per square meter	A/m^2
voltage difference	V	volt	V
conductivity	σ	siemens per meter	S/m
resistivity	ρ	ohm meter	Ω·m
Boltzmann's constant	k	joule per kelvin	J/K
Planck's constant	h	joule seconds	Js
Planck's constant divided by π	$\hbar$	joule seconds	Js
fundamental charge	e	coulomb	C
transmission coefficient	$\mathbf{T}$	ratio	

duction electrons:

$$\mathbf{v}_d = \frac{\sum_{i=1}^{n} \mathbf{v}_i}{n}. \tag{5.3}$$

The summation is taken over n conduction electrons per unit volume.

The equation of motion of a conduction electron in the presence of an electric field $\mathbf{E}$, now, becomes:

$$m\left(\frac{d\mathbf{v}_d}{dt} + \frac{\mathbf{v}_d}{\tau}\right) = (-e)\mathbf{E}. \tag{5.4}$$

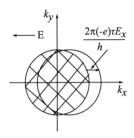

Fig. 5.1   The displacement of the Fermi sphere in the presence of an electric field **E**

The term proportional to the drift velocity represents the frictional force that resist the acceleration due to the field. The constant τ, which has a dimension of time, is called the **collision time** and can be seen as the average time since the last collision for an electron picked at random. In this formulation it is not specified with what the electron is colliding. When steady-state is reached $\mathbf{v}_d$ does not change, and $d\mathbf{v}_d/dt = 0$. The solution of equation 5.4 is immediately found as:

$$\mathbf{v}_d = \frac{(-e)\tau\mathbf{E}}{m}. \tag{5.5}$$

The Fermi sphere is thus displaced by the amount $\Delta k_x = (m/\hbar)\mathbf{v}_d = (-e)\tau E_x/\hbar$, when the electric field is applied, say, along the x-axis, see also figure Fig. 5.1.

Knowing that the electrical current density $\mathbf{J}$ is defined as $\mathbf{J} = n(-e)\mathbf{v}_d$, where n is the number of conduction electrons per unit volume, equation 5.5 is rewritten as:

$$\mathbf{J} = \left(\frac{ne^2\tau}{m}\right)\mathbf{E}. \tag{5.6}$$

Clearly, this equation predicts Ohm's law; in between the parenthesis there are only constants, and the conductivity is independent of the applied field. For an isotropic metal the electrical conductivity σ, defined by $\sigma = J/E$, is:

$$\sigma = \frac{ne^2\tau}{m}, \tag{5.7}$$

and the resistivity ρ is:

$$\rho = \frac{m}{ne^2\tau}. \tag{5.8}$$

Both τ and σ and ρ can be measured with experiments.

5.1.2 Electrical conductivity in quantum mechanics

There are two problems with the classical theory, especially when it is applied to metals. First, let us estimate the mean free path for Cu according to the Drude model. The mean free path Λ, that is, the average distance the electron travels between collisions is (assuming that the average velocity is much larger than the drift velocity):

$$\Lambda_{av} = v_{av}\tau. \tag{5.9}$$

The value of τ in pure Cu at room temperature is deduced using equation 5.8 using the observed resistivity of Cu from Table 5.2: $\rho = 1.55 \times 10^{-8}$ Ω-m. With the electron mass $m = 9.1 \times 10^{-31}$ kg, its charge $|e| = 1.6 \times 10^{-19}$ C, and the number density of atoms $n = 8.93$ g/m $\times 6.02 \times 10^{23}$ atoms per mol / 63.5 g per mol $= 8.47 \times 10^{28}$ atoms/m^{-3}, we find $\tau = 2.71 \times 10^{-14}$s.

Classically, the mean speed v_{av} is related to the thermal velocity defined by the relation (generally, at room temperature the random thermal velocity is much larger than the drift velocity):

$$v_{av} = \left(\frac{3kT}{m}\right)^{1/2} \approx 10^5 \text{ms}^{-1} \text{ at room temperature.} \tag{5.10}$$

Calculating the Λ_{av} we find that it is in the order of a few nanometers, however, the true mean free path of electrons is found to be as long as 20nm for pure Cu even at room temperature.

The second problem is that the classical theory gives also an incorrect temperature dependence for the resistivity. The mean free path Λ_{av} is temperature independent, it depends only on the density of lattice ions. According to Equation 5.10 the average velocity and thus the resistivity of

Table 5.2 Conductivity and resistivity of some metals at 273 K

Element	Electrical conductivity, σ ($\times 10^6$ /Ω-m)	Resistivity, ρ ($\mu\Omega$-cm)
Cu	64.5	1.55
Ag	66	1.5
Au	49	2.04
Al	40	2.50
Pb	5.7	19.3
Ti	2.38	42

typical metals is proportional to $\sqrt{T}$, provided that the number of electrons per unit volume n is temperature independent. To obtain a linear dependence one must assume that n has to change as $1/\sqrt{T}$, which is physically not acceptable.

The quantum mechanical description of the conduction considers the wave nature of the electron, it involves the scattering of electron waves by the lattice. Quantum mechanical calculations show that, for a *perfectly* ordered crystal at low temperatures, there is no scattering of electron waves. Therefore scattering occurs due to deviations from a perfectly ordered crystal. The most important deviations are thermal vibrations of the lattice ions or impurities in the . Electrons can also exchange energy with lattice vibrations. Another important result is that *only electrons at the very vicinity of the Fermi level* contribute to the current density. The full quantum mechanical calculation is based on the linearized Boltzmann transport equation and can be found in the physics literature.

We can, however, find the quantum mechanical expression of the conductivity by some semi-classical reasoning. Although the reasoning is not quite correct the answer is ok. Knowing that according to quantum mechanics the conduction electrons occupy energy levels near the Fermi level, we can use the expression of the number of free electrons per unit volume as a function of the number of conduction electrons at the Fermi level equation 4.79

$$n = \frac{2}{3} E_{\mathrm{F}} N(E_{\mathrm{F}}).$$

We can express this equation as

$$n = \frac{1}{3} m v_{\mathrm{F}}^2 N(E_{\mathrm{F}}). \tag{5.11}$$

If we substitute this expression for n in the formula for the conductivity

$$\sigma = \frac{n e^2 \tau}{m}$$

we find using $\tau = \Lambda_{\mathrm{F}}/v_{\mathrm{F}}$

$$\sigma = \frac{e^2}{3} \Lambda_{\mathrm{F}} v_{\mathrm{F}} N(\epsilon_{\mathrm{F}}), \tag{5.12}$$

Where Λ_{F} is the mean free path of the conduction electrons at the Fermi level and is defined by $\Lambda_{\mathrm{F}} = \tau v_{\mathrm{F}}$. Equation 5.12 shows that the conductivity is determined only by electrons at the Fermi level and is proportional to the number of electrons at the Fermi level $N(\epsilon_F)$, their velocity v_F, and the mean free path Λ_F. The Fermi velocity v_F is almost independent of the temperature, the collision time is also constant, but the mean free path Λ_F changes linearly with the temperature.

The equation holds for an electric field applied to a metal at a constant temperature (as a part of the linearized equation) and considers scattering of electron waves for which only deviations from a perfectly ordered lattice are important (the so-called relaxation time approximation). The equation, although, predicts that the conductance is almost independent of the temperature if the scattering of the conduction electron is considered to be elastic. Scattering with a static source of disturbances like impurity atoms or dislocations can be treated as being elastic, but scattering with lattice vibrations occurs through the exchange of energy with phonons. Detailed calculations of this mechanism show that these contributions give rise to the correct temperature dependence of the conductance at finite temperatures. A full quantum mechanical treatment can be found in [Mizutani (2001)], for example.

5.2 Current in Quantum Physics

The probability density of finding a particle $|\Psi(x,t)|^2$ is proportional to the probability of finding the particle described by the wave function $\Psi(x,t)$ in the interval dx around the point x, and represents a convenient starting point to deduce and charge currents in quantum mechanics. If the particle has charge q and is bound within some volume, we know that the total charge enclosed in that volume must be q, and the charge density $\rho(x)$ can be defined as:

$$\rho(x) = q|\Psi(x,t)|^2. \tag{5.13}$$

If we consider the charge density as a statistical quantity which is a continuous function of position, then the time derivative gives the rate of change of the charge density with time. In one dimension, we write down:

$$\frac{\partial \rho}{\partial t} = q\frac{\partial}{\partial t}|\Psi(x,t)|^2 = q\frac{\partial}{\partial t}[\Psi^*(x,t)\Psi(x,t)] = q(\Psi^*\frac{\partial \Psi}{\partial t} + \frac{\partial \Psi^*}{\partial t}\Psi). \tag{5.14}$$

The time rate of change of the charge density at any given point in space requires a difference between the particle currents flowing into and out of the differential volume surrounding the point in question to ensure the conservation of particles and charge. The mathematical statement of this fact is the continuity equation 3.14. Again, in one-dimension

$$\frac{\partial J}{\partial x} + \frac{\partial \rho}{\partial t} = 0. \tag{5.15}$$

in which J is the current density.

5.2.1   Current density in quantum physics

Equating the two expressions (5.14) and (5.15) for $\partial \rho / \partial t$ gives the relation (again in one-dimension)

$$\frac{\partial J}{\partial x} = -q(\Psi^* \frac{\partial \Psi}{\partial t} + \frac{\partial \Psi^*}{\partial t} \Psi) \tag{5.16}$$

which must be obeyed by the quantum mechanical analog of the current density J. To construct this current density we employ the time-dependent Schrödinger equation and its complex conjungate,

$$j\hbar \frac{\partial \Psi}{\partial t} = -\frac{\hbar^2}{2m} \frac{\partial^2}{\partial x^2} \Psi + U(x)\Psi \tag{5.17}$$

$$-j\hbar \frac{\partial \Psi^*}{\partial t} = -\frac{\hbar^2}{2m} \frac{\partial^2}{\partial x^2} \Psi^* + U(x)\Psi^* \tag{5.18}$$

In taking the complex conjungate, we have used the fact that $x = x^*$, $t = t^*$, and $U(x) = U(x)^*$ due to the fact that we are concerned with real positions, real times, and real potentials. Multiplying the Schrödinger equation by Ψ^* and its complex conjugate by Ψ to give

$$j\hbar \Psi^* \frac{\partial \Psi}{\partial t} = -\frac{\hbar^2}{2m} \Psi^* \frac{\partial^2}{\partial x^2} \Psi + \Psi^* U(x)\Psi \tag{5.19}$$

$$-j\hbar \Psi \frac{\partial \Psi^*}{\partial t} = -\frac{\hbar^2}{2m} \Psi \frac{\partial^2}{\partial x^2} \Psi)^* + \Psi U(x)\Psi^* \tag{5.20}$$

Subtracting the second from the first gives

$$j\hbar (\Psi^* \frac{\partial \Psi}{\partial t} + \Psi \frac{\partial \Psi^*}{\partial t}) = \frac{\hbar^2}{2m} (\Psi \frac{\partial^2}{\partial x^2} \Psi^* - \Psi^* \frac{\partial^2}{\partial x^2} \Psi) \tag{5.21}$$

The right-hand side can be simplified, using the rule of the derivative of a product:

$$\frac{\partial}{\partial x} (\Psi^* \frac{\partial}{\partial x} \Psi) = (\frac{\partial \Psi^*}{\partial x})(\frac{\partial \Psi}{\partial x}) + \Psi^* \frac{\partial^2}{\partial x^2} \Psi \tag{5.22}$$

When this is applied the products of the single derivatives cancel and the equation reduces to

$$j\hbar(\Psi^*\frac{\partial\Psi}{\partial t} + \Psi\frac{\partial\Psi^*}{\partial t}) = \frac{\hbar^2}{2m}\frac{\partial}{\partial x}(\Psi^*\frac{\partial}{\partial x}\Psi - \Psi\frac{\partial}{\partial x}\Psi^*) \qquad (5.23)$$

Substituting into Eq. 5.16 for $\partial J/\partial x$

$$\frac{\partial J}{\partial x} = \frac{\hbar q}{2jm}\frac{\partial}{\partial x}(\Psi^*\frac{\partial}{\partial x}\Psi - \Psi\frac{\partial}{\partial x}\Psi^*). \qquad (5.24)$$

Within an arbitrary constant, then,

$$J = \frac{\hbar q}{2jm}(\Psi^*\frac{\partial}{\partial x}\Psi - \Psi\frac{\partial}{\partial x}\Psi^*). \qquad (5.25)$$

The arbitrary constant is zero if $J = 0$ whenever $\Psi = 0$, as one would expect. Knowledge of the wave function Ψ therefore allows us to calculate the current density J quantum mechanically. For the case of electrons $q = $ e. In three dimensions the derivative $\partial\Psi/\partial x$ becomes the gradient $\nabla\Psi$.

The dependence on time vanishes from both ρ and J for a stationary state because $\exp(j\omega t)$ cancels between Ψ and Ψ^*. A stationary state may still carry a current (constant in time), however, a stationary state where $\Psi(x)$ is purely real carries no current. This applies to a particle in a box and to bound states in general. A superposition of bound states is then necessary to generate a current. This feature emphasizes the fact that the total wave function must be a complex quantity in general.

5.2.2 *Current of free electrons*

We consider a free electron moving in the $+x$-direction. It is described by $\Psi(x,t) = A\exp[j(kx - \omega t)]$. The quantum mechanical charge density is

$$\rho(x) = e|\Psi(x,t)|^2 = e|A|^2, \qquad (5.26)$$

uniformly over all space, and the current density is

$$J(x) = \frac{\hbar e}{2jm}(\Psi^*\frac{\partial}{\partial x}\Psi - \Psi\frac{\partial}{\partial x}\Psi^*) \qquad (5.27)$$

$$= \frac{\hbar e}{2jm}(jk|A|^2 + jk|A|^2) \qquad (5.28)$$

$$= \frac{\hbar e}{m}k|A|^2, \qquad (5.29)$$

or

$$J(x) = \frac{\rho \hbar k}{m} = \rho v, \tag{5.30}$$

where we used $\hbar k/m = p/m = v$. $J = \rho v$ is the expected result and is like the classical result $J = nev$.

The Schrödinger equation is linear, so further wave functions can be constructed by superposition of basic solutions. For example,

$$\Psi(x,t) = [A_+ \exp(jkx) + A_- \exp(-jkx)] \exp(-j\omega t) \tag{5.31}$$

describes a superposition of waves describing a free electrons travelling both directions. The quantum-mechanical expression for the current gives the expected result [Davies (1998)]

$$J = \frac{\hbar ek}{m}(|A_+|^2 - |A_-|^2). \tag{5.32}$$

5.2.3 *Purely real waves*

When the wave functions are real, the current density must be zero. The wavefunction and the complex conjugated wavefunction are the same, and thus the current density is zero. An interesting result is found for two counter-propagating *decaying* waves,

$$\Psi(x,t) = [B_+ \exp(\kappa x) + B_- \exp(-\kappa x)] \exp(-j\omega t) \tag{5.33}$$

Neither component would carry a current by itself because it is real, but the superposition gives

$$J = \frac{\hbar q \kappa}{jm}(B_+ B_-^* - B_+^* B_-) = \frac{2\hbar q \kappa}{m} \text{Im}(B_+ B_-^*). \tag{5.34}$$

The wave *must* contain components decaying in *both* directions, with a phase difference between them, for a current to flow. This imposes important constraints on the description of currents in nanostructures in terms of transmission and reflection of waves describing the electrons.

5.3 Tunneling and Tunnel Current

5.3.1 *Tunneling through a rectangular barrier*

To describe tunneling in a metal-insulator-metal structure consider a rectangular potential barrier. We are going to to determine the wave function

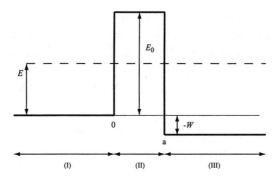

Fig. 5.2   Model for a potential energy barrier of width a and height E_0; E is the total energy of the electrons and W is the applied bias voltage.

$\psi(x)$ for a beam of free electrons, having a kinetic energy E_K, moving in a region in which the potential energy E_P is defined by:

$$E_P(x) = \begin{cases} 0 & (x < 0) & \text{region (I)} \\ E_0 & (0 \le x \le a) & \text{region (II)} \\ -W & (x > a) & \text{region (III)} \end{cases} \qquad (5.35)$$

as illustrated in figure Fig. 5.2.

The potential energy function consists of two steps in the potential energy. Defining E as the total energy, then for tunneling we consider the case $E < E_0$. Classical mechanics requires that the free electrons coming from the left with an energy $E < E_0$ should be reflected back at $x = 0$, because then their kinetic energies $E_K = E - E_0$ would be negative between $x = 0$ and $x = a$, which is impossible. Thus region (II) is a classical forbidden region if $E < E_0$.

Now, we consider the problem according to quantum mechanics by obtaining the solution of Schrödinger's equation for regions (I), (II), and (III). In region (I), in which $E_P = 0$, the Schrödinger equation becomes

$$\frac{\mathrm{d}^2\psi_1}{\mathrm{d}x^2} + \frac{2mE}{\hbar^2}\psi_1 = 0. \qquad (5.36)$$

which is identical to the Schrödinger equation for a free electron. Its general solution is of the type

$$\psi_1(x) = A e^{jkx} + B e^{-jkx}, \qquad (5.37)$$

with $k^2 = 2mE/\hbar^2$; the kinetic energy of the electrons $E_K = E$ because their potential energy is zero. In this way it is written it represents an incident electron beam (e^{jkx}) and a reflected electron beam (e^{-jkx}). We are assigning a different amplitude to the reflected electrons to take into account any possible change of the incident beam of electrons as a result of the reflection at $x = 0$.

In region (II), in which $E_P(x) = E_0$, Schrödinger's equation is

$$\frac{d^2\psi_2}{dx^2} + \frac{2m(E - E_0)}{\hbar^2}\psi_2 = 0. \qquad (5.38)$$

Because $E < E_0$, we define the positive quantity $\alpha^2 = 2m(E_0 - E)/\hbar^2$, so that the differential equation 5.38 becomes

$$\frac{d^2\psi_2}{dx^2} - \alpha^2\psi_2 = 0. \qquad (5.39)$$

The solution of this differential equation is a combination of the functions $e^{\alpha x}$ and $e^{-\alpha x}$, as may be verified by direct substitution. We must use both solutions because otherwise no current can flow through the barrier:

$$\psi_2(x) = Ce^{\alpha x} + De^{-\alpha x}. \qquad (5.40)$$

In region (III), in which $E_P = -W$, the solution is:

$$\psi_3(x) = Ge^{j\kappa x}, \qquad (5.41)$$

with $\kappa^2 = 2m(E - W)/\hbar^2$ because we only consider transmitted (tunnelled) electrons. The kinetic energy of the transmitted free electrons being $E_K = E - W$.

So, the wave function ψ_1 contains the incident and reflected electrons, ψ_2 decays exponentially, but it must also contain the positive exponential, since the barrier extends only up to $x = a$ and the positive exponential is necessarily to included a current. Because ψ_2 is not yet zero at $x = a$, the wave function continues into region (III) with the oscillating form ψ_3, which represents the transmitted electrons which have a total energy of $E - W$ (remember W is negative) with an amplitude G. Since ψ_3 is not zero, there is a finite probability of finding the electron in region (III). In other words, it is possible for an electron to go through the potential energy barrier even if its kinetic energy is less than the height of the barrier. Penetration of a potential energy barrier has no analog in classical mechanics because it correspond to a situation in which an electron has a negative kinetic energy or an imaginary momentum. We can do the calculation of the constants A, B, C, D, and G using the property that the wave function must be "well behaved".

- $\psi(x)$ is continuous at $x = 0$:

$$\psi_I(0) = \psi_{II}(0)$$
$$A + B = C + D \tag{5.42}$$

- $\frac{\partial \psi(x)}{\partial x}$ is continuous at $x = 0$:

$$\left.\frac{\partial \psi_I}{\partial x}\right|_{x=0} = \left.\frac{\partial \psi_{II}}{\partial x}\right|_{x=0}$$
$$jkAe^{jk0} - jkBe^{-jk0} = \alpha Ce^{\alpha 0} - \alpha De^{-\alpha 0}$$
$$jk(A - B) = \alpha(C - D) \tag{5.43}$$

- $\psi(x)$ is continuous at $x = a$:

$$\psi_{II}(a) = \psi_{III}(a)$$
$$Ce^{\alpha a} + De^{-\alpha a} = Ge^{j\kappa a} \tag{5.44}$$

- $\frac{\partial \psi(x)}{\partial x}$ is continuous at $x = a$:

$$\left.\frac{\partial \psi_{II}}{\partial x}\right|_{x=a} = \left.\frac{\partial \psi_{III}}{\partial x}\right|_{x=a}$$
$$\alpha(Ce^{\alpha a} - De^{-\alpha a}) = j\kappa Ge^{j\kappa a} \tag{5.45}$$

What the boundary conditions can give us are reflection and transmission probabilities . Such an approach is based on the assumption that a continuous incident beam of monoenergetic electrons gives rise to a continuous beam of transmitted electrons divided according to the same probabilities as the probability that a single electron in region(I) can be found in region(III). To find reflection and transmission probabilities, we need only to find ratios of numbers per unit time in the corresponding beams. Solving equations 5.42, 5.43, 5.44, and 5.45 gives [Fromhold (1981)] for the transmission probability $\mathbf{T}$:

$$\mathbf{T} = \frac{v'|G|^2}{v|A|^2} = \frac{4(\kappa/k)}{[1 + (\kappa/k)]^2 + [1 + (\kappa/\alpha)^2][1 + (\alpha/k)^2]\sinh^2 \alpha a} \tag{5.46}$$

In the limit that $W \to 0$, $\kappa \to k$ ($v' \to v$) and the transmission coefficient reduces to

$$\mathbf{T} = \frac{1}{1 + 1/4[(k/\alpha) + (\alpha/k)]^2 \sinh^2 \alpha a} \tag{5.47}$$

or

$$\mathbf{T} = \frac{4\frac{E}{E_0}\left(1 - \frac{E}{E_0}\right)}{\sinh^2\left[\sqrt{2m(E_0 - E)} \cdot a/\hbar\right] + 4\frac{E}{E_0}\left(1 - \frac{E}{E_0}\right)} \tag{5.48}$$

In general $\mathbf{T}$ is nonzero.

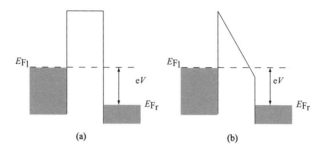

Fig. 5.3   A barrier between two metals as $T = 0$ K. The barrier is surrounded by a Fermi sea of free electrons. Different shapes of the barrier gives different transmission coefficients.

5.3.2   *Tunnel current*

In chapter 2 we saw that the tunnel current in the tunnel diode is given by

$$i = A \int_{E_c}^{E_v} \{f_c(E) - f_v(E)\} \mathbf{T}(E) \mathrm{d}E, \qquad (5.49)$$

if we assumed the density of states on both sites to be equal. In the same way we can discuss the tunnel current in case we have two leads of the same metal separated by a small gap at $T = 0$ K; see figure Fig. 5.3.

At zero kelvin the levels are filled completely upto the Fermi level. The difference between the Fermi level at the left side and the one at the right side is eV, where V is the (bias) voltage over the junction. Analogue to the tunnel current in the tunnel diode we have, assuming a uniform density of states:

$$i = B \int_{E_{\mathrm{Fr}}}^{E_{\mathrm{Fl}}} \mathbf{T}(E) \mathrm{d}E,$$

where B is a constant. That is

$$i = eV\mathbf{BT}(E). \qquad (5.50)$$

For small bias voltages the transmission coefficient belonging to barrier (a) in figure Fig. 5.3, given in equation 5.46, can considered to be constant. As a result we can expect a linear iv-relationship:

$$i = e\mathbf{BT}V = R_{\mathrm{t}}V,$$

R_{t} is called the tunnel resistance. Note that the Coulomb blockade phenomenon is not taken into account here. For larger bias voltages the transmission coefficient has an exponential behavior, which will dominate the iv-curve.

Sometimes it is better to model the barrier as in figure Fig. 5.3(b). This is especially true if the leads are separated by a vacuum. The triangular barrier must be calculated using a approximation technique called the WKB method. Tunneling in this case is called Fowler-Nordheim tunneling.

5.4 Shrinking Dimensions and Quantized Conductance

The continuing decrease in device dimensions will change the modeling of the devices. The three dimensional description of electronic devices must be replaced by a two- or even a one dimensional description. In these cases the electrons are confined to two- or one dimension. As soon as the confinement plays a role the energy will be quantised again leading to so-called (electric) subbands or modes. In modeling these systems some surprising results are obtained. We will also look briefly into systems in which the device dimensions are smaller than the mean free path of the electrons at the Fermi level.

5.4.1 *Two-dimensions*

Just as is the case in three dimensions the energy is quantized if the electrons are confined. We repeat the calculations of the free-electron model in subsection 4.5.4, but now in two dimensions. The energy quantization is:

$$E = \frac{\pi^2 \hbar^2}{2ma^2}(n_1^2 + n_2^2) \tag{5.51}$$

where n_1, and n_2 are integers. For a small sheet the energy levels are spaced widely. The quantum states can be represented in a two dimensional κ world ($\kappa^2 = n_1^2 + n_2^2$). We can find the number of quantum states $N_{2D}(E)$:

$$N_{2D}(E) = \frac{1}{4}\pi\kappa^2 = \frac{1}{2}\frac{a^2 m}{\pi\hbar^2}E. \tag{5.52}$$

The number of states with energy between E and $E + dE$ per unit area is obtained by differentiating the above equation and dividing it by a^2, yielding

$$dN_{2D}(E) = \frac{1}{2}\frac{m}{\pi\hbar^2}dE. \tag{5.53}$$

The density of states per spin direction per unit area, g_{2D}, is

$$g_{2\text{D}}(E) = \frac{1}{2}\frac{m}{\pi\hbar^2} = \frac{2\pi m}{h^2}. \tag{5.54}$$

We see that the density of states in each subband is the same. So the total density of states can be described by a stepping function and looks like a staircase with jumps at energies of the subbands. Modern examples of two dimensional electron gas systems are for example GaAs-AlGaAs heterostructures at low temperatures. But also, graphene nanoribbons (thin strips of graphene or unrolled single-walled nanotubes), or quantum point contacts (QPCs).

5.4.2   One-dimension and the quantum wire

If we consider the confinement of electrons in only one dimension, a so-called quantum wire, and consider the current through such a wire a very interesting result is obtained. We start with the simple one dimensional energy equation

$$E = \frac{\pi^2\hbar^2}{2ma^2}k^2. \tag{5.55}$$

For wires with a very small cross-section the energy levels are spaced widely. We find the number of quantum states $N_{1\text{D}}(E)$ per unit length:

$$N_{1\text{D}}(E) = \frac{1}{2}k = \frac{1}{2}\frac{1}{\pi\hbar}(2mE)^{1/2}. \tag{5.56}$$

The number of states with energy between E and $E + \text{d}E$ per unit length is obtained by differentiating the above equation, yielding

$$\text{d}N_{1\text{D}}(E) = \frac{(2m)^{1/2}}{4\pi\hbar}(E)^{-1/2}\text{d}E. \tag{5.57}$$

And the density of states per spin direction per unit length $g_{1\text{D}}$ is

$$g_{1\text{D}}(E) = \frac{(2m)^{1/2}}{4\pi\hbar}(E)^{-1/2} = \frac{1}{2}\frac{m}{\pi\hbar}(2mE)^{-1/2}. \tag{5.58}$$

Now, we consider the current through such a quantum wire. Therefore we apply an energy source that leads to a certain voltage across the quantum wire (in this case the quantum wire is often called a channel). Again, we assume that we can model this voltage drop V by

$$E_{\mathrm{Fl}} - E_{\mathrm{Fr}} = eV, \tag{5.59}$$

where E_{Fl} and E_{Fr} are the Fermi levels at different sides of the wire. The total current for one subband is determined by the electron density in the subband times the group velocity in this subband (including the factor 2 for two spin directions)

$$I_{1\mathrm{D}} = 2 \int_{E_{\mathrm{Fr}}}^{E_{\mathrm{Fl}}} e g_{1\mathrm{D}}(E) v_{\mathrm{g}}(E) \mathrm{d}E. \tag{5.60}$$

To simplify this expression, we recognize in the expression of $g_{1\mathrm{D}}(E)$ equation 5.58 the momentum $p = (2mE)^{1/2}$. That is we can write

$$g_{1\mathrm{D}}(E) = \frac{1}{2} \frac{m}{\pi \hbar} (p)^{-1} = \frac{1}{2} \frac{m}{\pi \hbar m v_{\mathrm{g}}(E)} = \frac{1}{2} \frac{1}{\pi \hbar v_{\mathrm{g}}(E)}, \tag{5.61}$$

with $v_{\mathrm{g}}(E)$ the group velocity of the electrons. For the 1D current in the subband one obtains:

$$I_{1\mathrm{D}} = \frac{e}{\pi \hbar} \int_{E_{\mathrm{Fr}}}^{E_{\mathrm{Fl}}} \mathrm{d}E = \frac{2e^2}{h} V. \tag{5.62}$$

We notice that the current in the subband is proportional to the voltage across the channel (quantum wire). The result holds for each subband, so we come to the conclusion that the resistance in a quantum wire is quantised. The 1D conductance is quantized in units of $2e^2/h$—approximately $(12.9 \ \mathrm{K}\Omega)^{-1}$. The conductance quantisation has been observed experimentally in short 1D channels in an AlGaAs/GaAs two dimensional electron gas.

5.4.3 *Ballistic Transport and the Landauer formula*

In the previous section we found that the 1D resistance is quantised and independent of the length of a short quantum wire. It is clear that this means that such a wire does not obey Ohm's law. The reason for this must be found in the fact that in the short wire no scattering takes place. That is the mean free path of the electrons, determined by the lattice defects, impurities, and phonon interaction, is larger than the channel (wire) length. Transport in small devices, in which electrons may travel from one lead to the other without encountering any scattering event from scatterers, is called **ballistic transport**. In ballistic transport the electrons are only scattered at the device boundaries; that is they are moving like billiard balls. Let us look into some details of this ballistic transport. Consider a uniform channel with a finite voltage across it. In a non ballistic device the electrons will be accelerated by the electric field but due to scattering an average velocity of the electrons can be defined and a current exist in the device. In case of a ballistic device due to the absence of scatterers electrons should be accelerated and a current cannot be defined. However, we can measure a current. This means that the electric field, and thus the voltage drop inside the central part of the channel have to be zero. Therefore, inside the channel, current flows without the existence of an electric field.

The effect of current flow without the presence of an electric field has been observed regardless the length and width of the channel as long as the channel is shorter than the mean free path. Nevertheless, due to energy sources to drive the current a voltage difference will present across the channel. The voltage can only drop at the leads. It is now assumed that a resistance arises because most of the electrons are backscattered upon arriving at the device and only a fraction of the electrons can enter the channel. This description leads to a model in which the conductance of the channel is described with transmission, T, and reflection, R, coefficients at the leads:

$$G = \frac{2e^2}{h}\frac{T}{R}, \tag{5.63}$$

the equation is know as the Landauer formula. As there will be subbands in the device, the channel is often compared with the model for a waveguide.

The picture of the electron in ballistic transport is that of a particle, and as such the mean free path must be longer than the channel length. In the description as a particle the mean free path is determined by lattice defects, phonon interactions, and impurities. At low temperatures in a semiconductor channel the impurities will dominate in the description. Sometimes the channel may to be longer. The description is then know as **coherent transport**. In the description of coherent transport the electron is described as a wave and only phase changing scattering is considered, and one considers the so-call phase coherence length. The difference with ballistic transport is due to the fact that elastic scattering at impurities does not destroy the phase memory of an electron. It is clear that a resistance due to coherent transport can not be described by Ohm's law too.

Problems and Exercises

5.1 Calculate the mean speed of free electrons at room temperature.

5.2 Derive equation 5.32.

5.3 Consider a tunneling junction between two aluminum leads. When the width of the tunnel barrier is 6 nm and the applied voltage is 0.2 V. consider only electrons tunneling from the Fermi level. In what order is their transmission probability? (Hint: use the work function of Al and the Fermi-energy level.)

5.4 Do a literature search and discuss modern devices exploiting ballistic transport.

5.5 What kind of circuit model would you use to model quantized resistance? Explain why.

Chapter 6

Energy in Circuit Theory

In this chapter we leave quantum physics and return to circuits and circuit theory. The effect of tunneling and Coulomb blockade will be modeled by (parasitic)lumped circuit elements. From a classical physics and electromagnetic theory point of view using Maxwell's equations, the **lumped circuit** is only an approximation as was discussed in chapter 3. It assumes steady-state currents or a quasi-static approach. Energy is conserved and if applied to circuits that cannot radiate, Kirchhoff's current and voltage laws could be obtained. In this chapter concepts of energy in circuit theory are briefly reviewed that play an important role in analyzing and synthesizing circuits with tunneling elements. Notations and definitions are introduced.

6.1 Lumped Circuits

For lumped circuits, the voltage *across* any element and the current *through* any element are always well-defined. In the theory describing lumped circuits, from now on called (electronic) circuit theory, the Kirchhoff laws enter axiomatically.

6.1.1 *Kirchhoff's laws*

The **topology** of the a circuit specifies the location of nodes and elements or more precise: nodes and branches; a **branch** may represent any two-terminal element, and the terminals are called **nodes**. In discussing circuits we describe the topology with nodes Ⓝ, branch (element) currents i_n, and branch (element) voltages v_n. Kirchhoff's current law (KCL) and Kirchhoff's voltage law (KVL) are the two fundamental *postulates* of circuit theory. They hold irrespective of the nature of the elements constituting the circuit. Both reflect topological properties of the circuit.

Based upon the fundamental law of physics that electric charge is conserved, **Kirchhoff's current law (KCL) for nodes** states that, for all times t, the algebraic sum of the currents $i_1, i_2, ..., i_n$ *leaving* any node is equal to zero:

$$\sum_n i_n(t) = 0. \tag{6.1}$$

Given any connected lumped circuit having n nodes, we may choose arbitrarily one of these nodes as a **datum node**, i.e., as a reference for measuring (electric) potentials $u_1, u_2, ..., u_{n-1}$. Let v_i denote the voltage (potential) difference across branch (element) i connected between node k and j. **Kirchhoff's voltage law (KVL)** states that, for all choices of datum node, for all times t, and for all pairs of nodes k and j:

$$v_i(t) = u_k(t) - u_j(t). \tag{6.2}$$

If we now consider a closed node sequence, thus starting end ending at the same node, the voltage difference must be zero. Now, Kirchhoff's voltage law (KVL) states that, for all closed node sequences (loops), for all times t, the algebraic sum of all node-to-node voltages (voltage differences) around the chosen closed node sequence is equal to zero:

$$\sum_n v_n(t) = 0. \tag{6.3}$$

The KCL and KVL always lead to *homogeneous linear algebraic* equations with *constant real coefficients*.

6.1.2   *Circuit elements*

A circuit is characterized by one or more *sources* of electrical energy interconnected with one or more receivers, or *sinks* of electrical energy. In fact, in the circuit energy is supplied by sources, transferred from one place in the circuit to another, and temporarily stored or dissipated in circuit elements. In general, for the circuit holds the principle of conservation of energy. The sum of the energy delivered by the independent sources in the circuit is equal to the energy absorbed by all other elements of the circuit (see subsection 6.2.1). Because the possibility of tunneling depends on the position of the Fermi energy levels, it is in our context useful to classify

Table 6.1   Important quantities and symbols used in this chapter

Quantity	Symbol	Unit	Symbol
time	t	second	s
charge	q	coulomb	C
current	I	ampère	A
voltage difference	v	volt	V
potential	u	volt	V
resistance	R	ohm	Ω
conductance	G	siemens	S
capacitance	C	farad	F
inductance	L	henry	H
conductivity	σ	siemens per meter	S/m
resistivity	ρ	ohm meter	$\Omega \cdot$m
energy	w	joule	J
power	P	joule per second	J/s

circuit elements to their ability to generate, to dissipate (to absorb), or to store energy.

To obtain a **subcircuit** we separate conceptually some part of circuit from its surroundings. In this way an entity is obtained between elements and the complete circuit; and an **element** is simply a subcircuit that cannot be decomposed any further. Since the total current entering an isolated circuit is zero, a subcircuit has at least two terminals, and the current entering the subcircuit through one terminal then leaves it through the other one. Since a voltage can only be defined, or measured, between two terminals, the description of a two-terminal subcircuit involves only one current—through the subcircuit—and one voltage—across the subcircuit.

A terminal pair is called a **port** whenever it is characterized by one current and one voltage. A subcircuit having only such terminal pairs is called an **n-port**. Within the context of this text, one-port elements will be considered mostly. *All* one-port elements will be represented schematically as shown in figure Fig 6.1; that is, both current and power are always directed towards the element.

For the analysis of the (equivalent) circuits, the following one-port elements are generally defined: linear and nonlinear *resistors* (R), linear and nonlinear *capacitors* (C), and linear and nonlinear *inductors* (L). The ideal energy generators *voltage source* (v_s) and *current source* (i_s) can be seen as nonlinear resistors. The elements are idealizations of the actual physical devices and relate the physical quantities *current* (i), *voltage* (v), *charge*

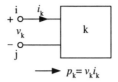

$$p_k = v_k i_k$$

Fig. 6.1   Schematic representation of a one-port network element k. The arrows follow the passive sign convention, so the current and the power are directed towards the element. Also, as a more general expression for the power $p_k(t) = \mathrm{Re}[v_k(t)i_k^*(t)]$ can be used, allowing complex variables.

(q), and *flux* $\phi(t)$. In circuit theory, resistors, capacitors, and inductors are abstractly defined, see for example [Chua *et al.* (1987)].

A one-port element is called a resistor if at any instant of time t its voltage $v(t)$ and its current $i(t)$ satisfy a *relation* F_R defined by a curve f_R in the iv-plane (or vi-plane), the curve may be time dependent:

$$F_R = \{(i,v) : f_R(i,v,t) = 0, \text{with } i = i(t) \text{ and } v = v(t)\} \qquad (6.4)$$

The curve is called the **characteristic of the resistor at time** t. It specifies the set of all possible values that the pair of variables $i(t)$ and $v(t)$ may take at time t. Usual physical resistors are time-invariant resistors; that is, its characteristic does not vary with time. A two-terminal subcircuit can behave like a time-varying resistor, if its characteristic in the iv-plane varies with time. For example, a switch can be considered a linear time-varying resistor. Any resistor can be classified in four ways depending upon whether it is linear or nonlinear and whether it is time-varying or time-invariant. A resistor is called a **linear resistor** if its characteristic is at all times a straight line through the origin. Accordingly, a resistor that is not linear is called a **nonlinear resistor**.

A **linear time-invariant resistor**, by definition, has a characteristic that does not vary with time and is a straight line through the origin. The relation between its instantaneous voltage $v(t)$ and current $i(t)$ is expressed by Ohm's law:

$$F_R = \{(i(t), v(t)) : v(t) - Ri(t) = 0\} \qquad (6.5)$$

or

$$F_R = \{(v(t), i(t)) : i(t) - Gv(t) = 0\} \qquad (6.6)$$

R and G are constants independent of i,v, and t. R is called the **resistance** and G is called the **conductance**. The two two-terminal elements **open**

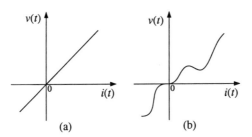

Fig. 6.2 Characteristics of: (a) (passive) time-invariant linear resistive element, (b) (passive) time-invariant nonlinear resistive element

circuit and **short circuit** can be considered to be linear time-invariant resistors with conductance zero and resistance zero, respectively. Figure Fig. 6.2 shows a (passive) time-invariant linear resistor and a (passive) time-invariant nonlinear resistor.

In the same way capacitors and inductors are defined. A two-terminal element is called a **capacitor** if at any time t its stored charge $q(t)$ and its voltage $v(t)$ satisfy a relation defined by a curve in the vq-plane:

$$F_\mathrm{C} = \{(v, q) : f_\mathrm{C}(v, q, t) = 0, \text{with } v = v(t) \text{ and } q = q(t)\} \qquad (6.7)$$

This curve is called the **characteristic of the capacitor at time** t. It specifies the set of all possible values that the pair of variables $v(t)$ and $q(t)$ may take at time t. As in the case of the resistor, a capacitor whose characteristics is at all times a straight line through the origin of the vq-plane is called a **linear capacitor**. Conversely, if at any time the characteristic is not a straight line through the origin of the vq-plane, the capacitor is called a **nonlinear capacitor**. A capacitor whose characteristic does not change with time is called a time-invariant capacitor. If the characteristic does change with time, the capacitor is called a time-varying capacitor.

A **linear time-invariant capacitor**, by definition, has a characteristic that does not vary with time and is a straight line through the origin. The relation between its instantaneous voltage $v(t)$ and stored charge $q(t)$ is expressed by:

$$F_\mathrm{C} = \{(v(t), q(t)) : q(t) - Cv(t) = 0\} \qquad (6.8)$$

where C is a constant independent of q, v, and t. C is called the **capacitance**. The equation relating the current through and the terminal voltage across the capacitor is:

$$i(t) = \frac{dq}{dt} = C\frac{dv}{dt} \tag{6.9}$$

By integration of this equation the terminal voltage as a function of the current can be obtained within an arbitrary integration constant:

$$v(t) = \frac{1}{C}\int_{-\infty}^{t} i(\alpha)d\alpha = v(t_0) + \frac{1}{C}\int_{t_0}^{t} i(\alpha)d\alpha \tag{6.10}$$

and the supplementary datum $v(t_0)$, the **initial voltage**, must be specified in order to define the state of a capacitance.

The inductor is defined in a similar way. A two-terminal element is called a **inductor** if at any time t its flux $\phi(t)$ and its current $i(t)$ satisfy a relation defined by a curve in the $i\phi$-plane:

$$F_{\rm L} = \{(i,\phi) : f_{\rm L}(i,\phi,t) = 0, \text{with } i = i(t) \text{ and } \phi = \phi(t)\} \tag{6.11}$$

This curve is called the **characteristic of the inductor at time** t. It specifies the set of all possible values that the pair of variables $i(t)$ and $\phi(t)$ may take at time t. As in the case of the capacitor, an inductor whose characteristics is at all times a straight line through the origin of the $i\phi$-plane is called a **linear inductor**. Conversely, if at any time the characteristic is not a straight line through the origin of the $i\phi$-plane, the inductor is called a **nonlinear inductor**. An inductor whose characteristic does not change with time is called a time-invariant inductor. If the characteristic does change with time, the inductor is called a time-varying inductor.

A **linear time-invariant inductor**, by definition, has a characteristic that does not vary with time and is a straight line through the origin. The relation between its instantaneous current $i(t)$ and flux $\phi(t)$ is expressed by:

$$F_{\rm L} = \{(i(t),\phi(t)) : \phi(t) - Li(t) = 0\} \tag{6.12}$$

where L is a constant independent of ϕ,i, and t. L is called the **inductance**. The equation relating the terminal voltage and the current through the inductor is:

$$v(t) = \frac{d\phi}{dt} = L\frac{di}{dt} \tag{6.13}$$

By integration we get:

$$i(t) = i(t_0) + \frac{1}{L}\int_{t_0}^{t} v(\alpha)d\alpha \tag{6.14}$$

and the initial value $i(t_0)$ must be specified in order to define the state of the inductor.

6.1.3 *Energy considerations: passive and active elements*

For any continuous current i, a (nonlinear) resistor element with the product iv always greater than zero dissipates power, and is called **passive**. Otherwise the resistor is called **active**. Consequently, the characteristic of a passive, linear or nonlinear, resistive element must be confined to the first and third quadrant of the iv-plane, an example was shown in figure Fig. 6.2. However, the element may possess regions of negative slope, as can be seen in the figure, and still be passive.

For a passive linear resistor the **instantaneous power** defined as the product $p = vi = Ri^2$ is always nonnegative. If a resistive element with $R > 0$ dissipates power, then power is directed towards the element and we define the positive current reference going *into* the resistive element at the terminal with the positive voltage reference; this choice is generally called the **passive sign convention** (also see figure Fig. 6.1).

Active resistors have at least a part of their characteristic in the second or fourth quadrant of the iv-plane. Examples are the affine linear resistors and the energy generators: voltage source and current source.

Within an energy context it is useful to treat sources as separate one-port elements. A **voltage source**, with value eV, is a one-port defined by $v = e$, where e is a given function of time, see figure Fig. 6.3 (c,e). Taking into account the passive sign convention, if the current i or the voltage v is negative the source is generating energy; if the product of the current and voltage iv is positive the source is absorbing energy (also see figure Fig. 6.4).

A **current source**, with value jA, is a one-port defined by $i = (-j)$, where j is a given function of time, see figure Fig. 6.3 (d,f). Again, taking into account the passive sign convention, if the current i is negative, and thus $(-j)$ is negative, the source is generating energy; if the current i is positive the source is absorbing energy (figure Fig. 6.4).

Finally, the equation $v = 0$ defines a **short circuit**, which can also be considered as a zero-valued resistance. Similar, the equation $i = 0$ defines an **open circuit**, which can be considered as an zero-valued conductance.

For the description of charge transport *within* an island in nanoelectronics it is useful to model this charge transport explicitly by a circuit element. The element is an example of what is called a **nonenergetic element**. This element neither absorbs nor delivers power (energy). The element is best described as a short circuit, it's characteristic is defined as a line through the i-axis in the iv-plane. Because the element allows any

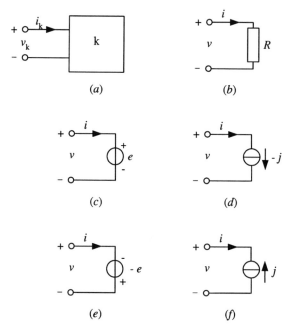

Fig. 6.3   (a) one-port element k with passive sign convention; (b) resistor element with value R (e.g., $R = 1$ kΩ); (c,e) voltage source element with value e (e.g., $e = 8$ V); (d,f) current source element with value j (e.g., $j = 5$ A)

current to flow without a voltage difference, it defines only charge transport per unit of time along the short circuit due to a current elsewhere created in the circuit. Figure Fig. 6.5 shows the short-circuit element, it's symbol, it's defining characteristic. As is clear the the short circuit element can also be seen as a subset of either the voltage source element (having value $e = 0$) or the resistor element (having value $R = 0$).

In addition to the dissipated power, we can consider power stored in capacitive and inductive elements. The electric energy stored in a linear capacitor w_{se} and the magnetic energy stored in a linear inductor w_{sm} are

$$w_{se} = \frac{1}{2}Cv^2; \qquad w_{sm} = \frac{1}{2}Li^2 \qquad (6.15)$$

respectively. Because the stored energy is always positive one has $C \geq 0$ and $L \geq 0$. More general, one might say that the characteristics of the capacitor and inductor are always monotonically increasing.

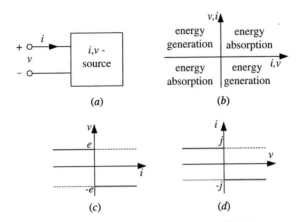

(a)   *(b)*

(c)   *(d)*

Fig. 6.4 Energy generation and absorption of the energy generating elements: voltage and current source. The solid line characteristics in (c) and (d) indicate energy generating, the dotted line characteristics indicate energy absorption.

6.1.4 *Linear elements and superposition*

Any lumped circuit with the property that each of its elements is either a *linear element* or an *independent source* is called a **linear circuit**. Important circuit theorems apply to linear networks. One of them is the superposition theorem. The **superposition theorem** can be defined as follows: *If the response of a system to a linear combination of input signals always consists of the corresponding combination of the individual output signals, then for this system the superposition principle applies.* That is, if a

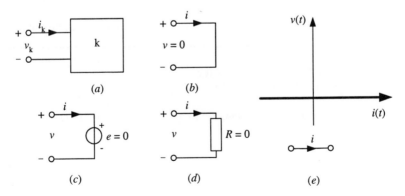

Fig. 6.5 The nonenergetic short-circuit element (a); (b, c, d) implementations of the element; (e) symbol and defining characteristic.

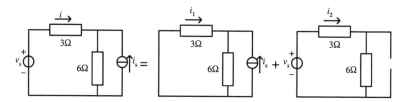

Fig. 6.6   Using the superposition principle it holds true that: $i = i_1 + i_2$

system responds to an input signal x_1 with the output signal y_1 and to the input signal x_2 with the output signal y_2 then the superposition theorem holds if:

$$x(t) = Ax_1(t) + Bx_2(t) \rightarrow y(t) = Ay_1(t) + By_2(t). \qquad (6.16)$$

A and B can be any complex constants. For the analysis and design of (nanoelectronic) circuits this theorem is of great importance. It allows us to decompose an original, more complicated, circuit analysis problem into a number of simpler ones by considering partial responses.

As an example, consider figure Fig. 6.6. The current i in the linear circuit can be found as the sum of the partial responses i_1 and i_2. In words, if i_1 is the response with the voltage source **deactivated** (reduced to zero, thus a short circuit) and i_2 is the response with the current source deactivated (reduced to zero, thus an open circuit) then $i = i_1 + i_2$. We see that: $i_1 = -(2/3)i_s$ and $i_2 = (1/9)\Omega^{-1}v_s$, so $i = (1/9)\Omega^{-1}v_s - (2/3)i_s$.

Superposition *cannot* be used to calculate dissipated energy of individual elements (for example, you can see this realizing that: $((1/9)\Omega^{-1}v_s - (2/3)i_s)^2 \neq ((1/9)\Omega^{-1}v_s)^2 + (-(2/3)i_s)^2)$.

Using linear resistors linear voltage and current division can be obtained.

Example 6.1 As an example, it is proven that the superposition theorem applies for a linear time-invariant resistor:
▽ If $v_1(t) = Ri_1(t)$ and $v_2(t) = Ri_2(t)$ then

$$v(t) = v_1(t) + v_2(t) \text{ must hold true if } i(t) = i_1(t) + i_2(t).$$

Starting with the currents, we obtain easily:

$$i_1(t) = \frac{1}{R}v_1(t), \qquad i_2(t) = \frac{1}{R}v_2(t)$$

$$i(t) = i_1(t) + i_2(t) = \frac{1}{R}(v_1(t) + v_2(t))$$

$$v(t) = Ri(t) = \frac{R}{R}(v_1(t) + v_2(t)) = v_1(t) + v_2(t),$$

which completes the proof.

△

Example 6.2 As an example, it is proven that with two linear resistors a linear voltage division can be obtained, see figure Fig. 6.7:

▽ For the circuit holds:

$$v(t) = v_1(t) + v_2(t), \quad i(t) = i_1(t) = i_2(t).$$

$$v_1(t) = R_1 i_1(t), \quad \text{and} \quad v_2(t) = R_2 i_2(t)$$

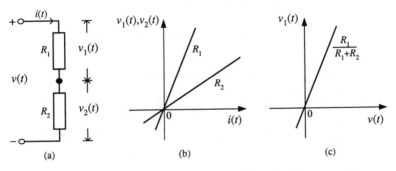

Fig. 6.7 Linear voltage division: (a) subcircuit, (b) definition of the linear resistors, and (c) linear voltage division

We obtain easily:

$$v(t) = R_1 i(t) + R_2 i(t) = (R_1 + R_2)i(t),$$

and

$$v_1(t) = R_1 i(t) = \frac{R_1}{R_1 + R_2} v(t).$$

So the voltage transfer characteristic is linear, which completes the proof.

△

Other network theorems, such as the Thévenin (Norton) equivalent network theorems, apply to linear circuits only too. For our purposes, the superposition theorem can be used to find out whether the circuit under consideration is a linear circuit. If equation 6.16 does not hold the circuit is nonlinear. Knowing that a circuit is a linear circuit, and time-invariant,

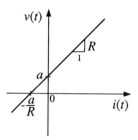

Fig. 6.8   Characteristics of an affine linear resistor

most circuit theorems and the method of Laplace transforms can be used as effective tools in analyzing the circuits.

Linear circuits may include linear resistors, linear inductors, linear capacitors, linear dependent sources and independent sources.

6.1.5   *Affine linear and nonlinear elements*

Any lumped circuit that is not a linear circuit is called a **nonlinear** circuit. The most important observation is that, in general, superposition does *not* apply to nonlinear circuits. A special class of nonlinear elements are the so-called, **affine linear elements**, which are characterized by a straight line *not* passing through the origin.

For example, an affine linear resistor is defined by, see also figure Fig. 6.8:

$$F_R = \{(i(t), v(t)) : v(t) - Ri(t) = a, \text{with } a \neq 0\} \tag{6.17}$$

Example 6.3 As an example, it is proven that the superposition theorem does not apply for an affine linear time-invariant resistor. It is shown that if superposition is assumed a contradiction will follow:

▽ If $v_1(t) = Ri_1(t) + a$, $(a \neq 0)$ and $v_2(t) = Ri_2(t) + a$, $(a \neq 0)$ then

$v(t) = v_1(t) + v_2(t)$ must hold true if $i(t) = i_1(t) + i_2(t)$.

Starting with the currents, we obtain:

$$i_1(t) = \frac{1}{R}(v_1(t) - a), \qquad i_2(t) = \frac{1}{R}(v_2(t) - a)$$

$$i(t) = i_1(t) + i_2(t) = \frac{1}{R}(v_1(t) + v_2(t) - 2a)$$

$$v(t) = Ri(t) + a = \frac{R}{R}(v_1(t) + v_2(t) - 2a) + a =$$
$$= v_1(t) + v_2(t) - a \neq v_1(t) + v_2(t),$$

which completes the proof.

$\triangle$

An affine linear voltage division will be obtained if we consider a linear resistor in series with an affine linear resistor, or in case we have two affine linear resistors (except for two affine resistors in which $a_1 = -a_2$) .

Example 6.4 As an example, it is proven that an affine linear resistor in series with a linear resistor leads to an affine voltage division, see Fig. 6.9.

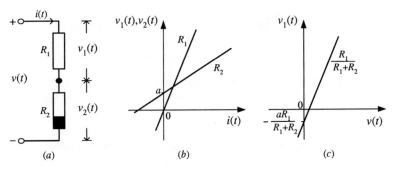

Fig. 6.9 Affine linear voltage division: (a) subcircuit, (b) definition of the resistors, and (c) affine linear voltage division

$\triangledown$ For the circuit holds:

$$v(t) = v_1(t) + v_2(t), \qquad i(t) = i_1(t) = i_2(t).$$

$$v_1(t) = R_1 i_1(t), \quad \text{and} \quad v_2(t) = R_2 i_2(t) + a.$$

So,

$$v(t) = R_1 i(t) + R_2 i(t) + a$$

and thus

$$i(t) = \frac{1}{R_1 + R_2} v(t) - \frac{a}{R_1 + R_2}.$$

For $v_1(t)$ and $a \neq 0$ we obtain:

$$v_1(t) = R_1 i(t) = \frac{R_1}{R_1 + R_2} v(t) - \frac{aR_1}{R_1 + R_2}.$$

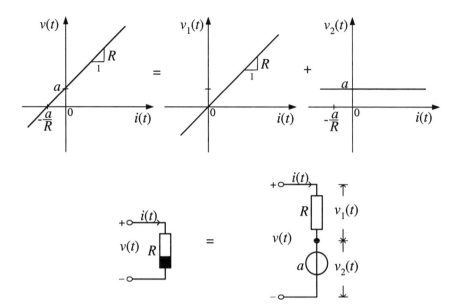

Fig. 6.10   Decomposition of an affine linear resistor, by splitting voltages ($v(t) = v_1(t) + v_2(t)$), into a series combination of a linear resistor and a voltage source; the resulting circuit is a linear circuit.

So the voltage transfer characteristic is affine linear, which completes the proof.
△

However, if there is a single nonlinear element in a circuit, we can often choose two independent sources with appropriate waveforms so that superposition does hold. Particularly, this holds for affine linear resistors.

In figures Figs. 6.10 and 6.11 the transformations into equivalent subcircuits are done. Of course, one recognizes the two equivalent circuits as Thévenin/Norton equivalents (see also the discussions in the next section).

In summary, in nonlinear circuits the superposition theorem does not apply. This will, in general, complicate circuit analysis of those circuits. In chapter 1, it was shown that in Coulomb blockade two SET junctions in series including a non-zero island charge predict an affine linear voltage transfer characteristic. In this section we have seen that such a characteristic can also be obtained by the series combination of a affine linear resistor and a linear one, or by a series combination of two affine linear resistors. We also saw that, in case of an affine linear resistor the element can be transformed to a linear equivalent subcircuit by introducing suit-

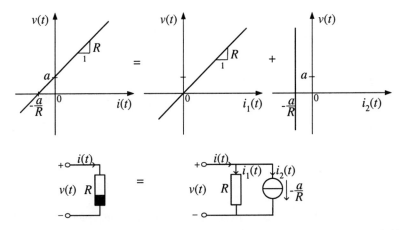

Fig. 6.11 Decomposition of an affine linear resistor, by splitting currents $(i(t) = i_1(t) + i_2(t))$, into a parallel combination of a linear resistor and a current source; the resulting circuit is a linear circuit.

able energy sources. This possibility greatly simplifies the analysis of such circuits. In the next chapter we will investigate the possibility to perform such a transformation on the two SET junctions while studying the island charge.

6.2   Circuit Theorems

The best approach to finding a circuit element, or subcircuit, for nanoelectronic devices is based on treating them as "black boxes", that is, subcircuits in which the specific behavior such as tunneling is hidden from the outside. If these elements can be found then general circuit theorems on these black boxes can reveal the possibilities and restrictions that a circuit theory imposes on them. In the following we discuss Tellegen's theorem and the Thévenin and Norton equivalents.

6.2.1   *Tellegen's theorem*

In all the circuits and networks that were considered upto here energy was conserved. This is no coincidence, it is a general property of lumped circuits and is based on a theorem called Tellegen's theorem. Tellegen's theorem is general in that it depends solely upon Kirchhoff's laws and the topology of the network. It applies to all (sub)circuits and elements, whether they

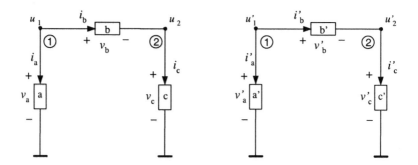

Fig. 6.12   Example of the Tellegen's Theorem

be linear or nonlinear, time-invariant or time variant, passive or active, hysteretic or nonhysteretic. The excitation might be arbitrary; the initial conditions may also be arbitrary. It describes elements as "black boxes" only assuming a circuit topology and the validity of Kirchhoff's laws.

-Tellengen's theorem

Tellegen's theorem (1952) can be formulated as follows. Suppose we have *two* circuits with the same topologies, and all N elements are represented by a one-port as in figure Fig. 6.1. For each element k in one circuit (the unprimed circuit), there is a corresponding element k' in the other circuit (the primed circuit). If we now take the *voltage* across element k, v_k, and multiply it with the *current* through the corresponding element k', i'_k, and sum this for all N elements then:

$$\sum_{k=1}^{N} v_k i'_k = 0. \qquad (6.18)$$

To fully understand this theorem consider the next example [Davis (1998)], see figure Fig. 6.12.

We have a unprimed circuit and a primed circuit consisting of three elements a, b, and c. Using Kirchhoff's voltage law the voltages v across the elements are expressed in potentials u with respect to the datum node (ground). Now we can easily check that:

$$v_a i'_a + v_b i'_b + v_c i'_c = u_1 i'_a + (u_1 - u_2)i'_b + u_2 i'_c =$$
$$u_1(i'_a + i'_b) + u_2(-i'_b + i'_c) = u_1 0 + u_2 0 = 0,$$

because according to Kirchhoff's current law the currents at node ① and ② sum up to zero.

-Proof of Tellegen's theorem

Consider a general element in a given circuit—the unprimed circuit—and a corresponding element in another circuit—the primed circuit—having exactly the same topology. Because the topologies are the same, there is for any element between nodes ⓐ and ⑧ in the unprimed circuit a corresponding element between the same nodes in the primed circuit. We assume that there are N elements and n_t nodes and investigate the sum

$$\sum_{k=1}^{N} v_k i'_k = 0.$$

Let us assume, for simplicity, that the circuit has no branches in parallel; i.e., there exists only one branch between any two nodes. The proof can be easily extended to the general case. (If branches are in parallel, replace them by a single branch whose current is the sum of the branch currents. If there are several parts, the proof shows that Tellegen's theorem holds for each of them. Hence it holds also when the sum ranges over all branches of the graph.)

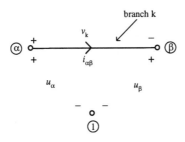

Fig. 6.13 An arbitrary branch, k, connects node ⓐ and node ⑧

We first pick an arbitrary node as a reference node, and we label it node ①. Thus, $u_1 = 0$. Let u_α and u_β be potentials of the αth node and the βth node, respectively, with respect to the reference node. It is important to note—in both the unprimed circuit and the primed circuit—that once the branch voltages $(v_1, v_2, ..., v_N)$ are chosen, then by KVL the node potentials $u_1, u_2, ..., u_\alpha, ..., u_\beta, ...$ are uniquely specified. Let us assume that branch k connects node ⓐ and node ⑧ as shown in Figure 6.13, and let us denote by $i'_{\alpha\beta}$ the current flowing in branch k *from* node ⓐ *to* node ⑧ in the primed circuit. The voltage v_k is the voltage across element k in the unprimed circuit, then

$$v_k i'_k = (u_\alpha - u_\beta) i'_{\alpha\beta} \tag{6.19}$$

Obviously, $v_k i'_k$ can also be written in terms of $i'_{\beta\alpha}$, the current *from* node ⓑ *to* node ⓐ in the primed circuit , as

$$v_k i'_k = (u_\beta - u_\alpha) i'_{\beta\alpha}$$

Adding the above two equations, we obtain

$$v_k i'_k = \frac{1}{2} \left[(u_\alpha - u_\beta) i'_{\alpha\beta} + (u_\beta - u_\alpha) i'_{\beta\alpha} \right] \tag{6.20}$$

Now, if we sum the left-hand side of equation 6.20 for all the branches in the graph we obtain

$$\sum_{k=1}^{N} v_k i'_k$$

The corresponding sum of the right-hand side of equation 6.20 becomes, using $i'_{\beta\alpha} = -i'_{\alpha\beta}$,

$$\frac{1}{2} \sum_{\alpha=1}^{n_t} \sum_{\beta=1}^{n_t} (u_\alpha - u_\beta) i'_{\alpha\beta}$$

where the double summation has indices α and β carried over all the nodes in the graph. The sum leads to the following equation:

$$\sum_{k=1}^{N} v_k i'_k = \frac{1}{2} \sum_{\alpha=1}^{n_t} \sum_{\beta=1}^{n_t} (u_\alpha - u_\beta) i'_{\alpha\beta}. \tag{6.21}$$

Note that if there is no branch joining node ⓐ to node ⓑ we set $i'_{\alpha\beta} = i'_{\beta\alpha} = 0$. Now that equation 6.21 has been established, the right-hand side of equation 6.21 is split as follows:

$$\sum_{k=1}^{N} v_k i'_k = \frac{1}{2} \sum_{\alpha=1}^{n_t} u_\alpha \left(\sum_{\beta=1}^{n_t} i'_{\alpha\beta} \right) - \frac{1}{2} \sum_{\beta=1}^{n_t} u_\beta \left(\sum_{\alpha=1}^{n_t} i'_{\alpha\beta} \right). \tag{6.22}$$

For each fixed α, $\sum_{\beta=1}^{n_t} i'_{\alpha\beta}$ is the sum of *all* branch currents entering node ⓑ. By KCL, each one of these sums is zero, hence,

$$\sum_{k=1}^{N} v_k i'_k = 0.$$

Thus we have shown that given any set of branch voltages subject to KVL only and any set of branch currents subject to KCL only, the sum of the products $v_k i'_k$ is zero. This concludes the proof.

- *Conservation of energy in lumped circuits*

For a single arbitrary circuit we can take the same circuit for both the primed circuit and the unprimed circuit. Tellegen's theorem now states the conservation of energy in the circuit. We have, with the notations of Tellegen's theorem,

$$\sum_{k=1}^{N} v_k(t) i_k(t) = 0 \qquad \text{for all } t.$$

Since $v_k(t) i_k(t)$ is the power delivered at time t by the circuit to branch k, the theorem may be interpreted as follows: at any time t the sum of the power delivered to each branch of the circuit is zero. Suppose the circuit has several independent sources; separating in the sum the sources from the other branches, we conclude that *the sum of the power delivered by the independent sources to the circuit is equal to the sum of the power absorbed by all the other elements of the circuit.* This means that as far as lumped circuits are concerned, the validity of KVL and KCL together imply conservation of energy.

6.2.2 *Thévenin and Norton equivalents*

The application of the superposition principle leads to the important linear Norton and Thévenin theorems that are frequently used in linear circuit analysis to obtain Norton and Thévenin equivalent subcircuits. In case of nonlinear circuits it is also possible to define Norton and Thévenin equivalent subcircuits but, in general, the resistances in the Norton and Thévenin equivalents will be different.

The linear theorems state that for most complex (passive) circuits there exists an equivalent circuit consisting of just one resistance and one independent source. The **Thévenin equivalent** consists of a resistance R_{Th} in series with an independent voltage source $v_{oc}(t)$. The voltage is called the open-circuit voltage, and has the value of the voltage that appears across the two terminals when no other circuitry is attached. The resistance R_{Th} is called the **Thévenin equivalent resistance** and has the value of the equivalent resistance when all independent sources are deactivated.

The **Norton equivalent** consists of a resistance R_{Th} in parallel with an independent current source $i_{sc}(t)$, see figure Fig. 6.14. The current is

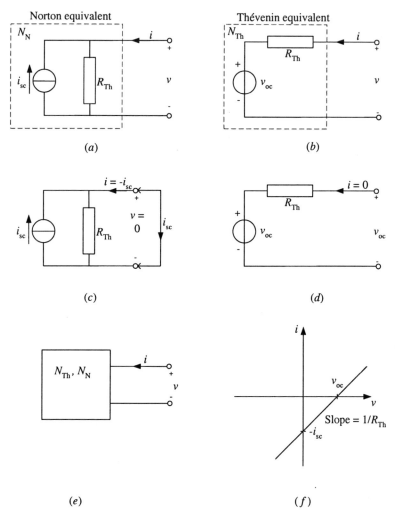

Fig. 6.14 Definitions used in the linear passive Norton and Thévenin equivalents: (a) the Norton equivalent source model, (b) the Thévenin equivalent source model, (c) the short circuit current to obtain the Norton equivalent, (d) the open circuit to obtain the Thévenin equivalent, (e) the equivalent circuits as an element, and (f) the characteristic in the vi-plane for these models.

called the short-circuit current, and has the value of the current that flows through a short circuit of the two terminals. The resistance R_{Th} is again called the Thévenin equivalent resistance and has the value of the equivalent resistance when all independent sources are deactivated These subcircuit concepts are well-known, for definitions see figure Fig. 6.14.

Notice that it is not necessary to make assumptions concerning the load; it may be linear, nonlinear, time-variant, or time-invariant. The network N is only required to be linear; it may include both dependent and independent sources.

Within our energy content an important remark must be made. As the equivalent subcircuits are *only* equivalent to the original subcircuit because they have the same characteristics in the iv-plane, they do not say much about absorbed or generated power *in* the circuits. Generally, energy consumption is *not* the same in the subcircuit and its Thévenin or Norton equivalent.

Example 6.5 The following example shows that the Thévenin equivalent *cannot* be used to calculate power consumption within the circuit. Figure Fig. 6.15 shows a circuit, consisting of a subcircuit N and a load resistance, and the simplified circuits using the Thévenin equivalent of the subcircuit N_{Th}, and using the Norton equivalent of the subcircuit N_{N}. We compute the power loss in both circuit and equivalent subcircuits and compare them.

▽ The power loss in subcircuit N, figure Fig. 6.15(a) is:

$$P_N = \sum i^2 R = (2A)^2 (2\Omega) + (1A)^2 (2\Omega) = 10W.$$

The power loss in subcircuit N_{Th}, figure Fig. 6.15(b) is:

$$P_{N_{Th}} = \sum i^2 R = (1A)^2 (1\Omega) = 1W.$$

The power loss in subcircuit N_{N}, Fig. 6.15(c) is:

$$P_{N_N} = \sum i^2 R = (2A)^2 (1\Omega) = 4W.$$

Of course, the total energy absorbed by all resistors in both circuit (a), (b), and (c) is exactly the amount of energy provided by the source in the circuits: 12W, 3W, and 6W, respectively.

△

Problems and Exercises

6.1 Consider the vi-characteristic of the tunnel diode. Characterize this nonlinear resistor.

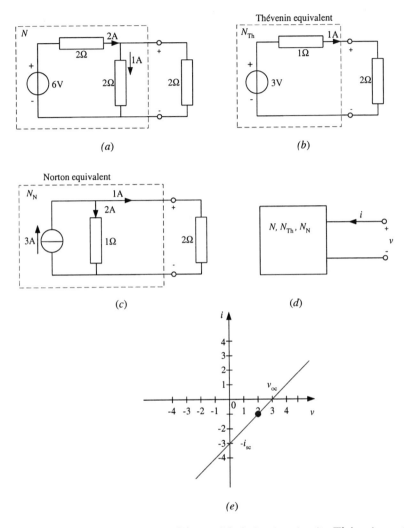

Fig. 6.15 (a) the circuit example, (b) simplified circuit using its Thévenin equivalent, (c) simplified circuit using its Norton equivalent, and (e) the characteristic in the vi-plane. The dot indicate the point on the characteristic for the circuit with the 2Ω load attached to it.

6.2 Determine the voltages across and the current through the linear resistors for each of the three circuits in figure Fig. 6.16.

6.3 For the three circuits in figure Fig. 6.16 calculate the power dissipated

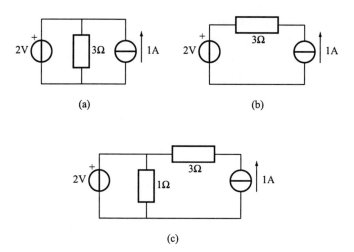

Fig. 6.16

in each resistor. Determine where this power comes from by calculating the contribution due to the voltage source and that due to the current source.

6.4 For the two networks of figure Fig. 6.17. (a) Find the currents through and the voltages across the elements. (b) Verify Tellegen's Theorem by taking the voltages of N_1 and the currents of N_2.

6.5 Verify Tellegen's Theorem by taking the voltages at $t = 0$ of N_1 and the currents of N_2 at $t = 10$ s of figure Fig. 6.17.

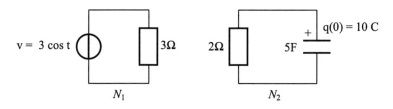

Fig. 6.17

6.6 Perform a circuit simulation of the circuit in figure Fig. 6.18. (a) Write down the values of voltages across and currents through all elements.

(b) Change *all* values of the elements, perform a new circuit simulation. Write down again the values of voltages across and currents through all elements. (c) Verify Tellegen's theorem by taking all currents of the first simulation and the voltages of the second.

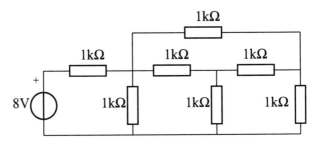

Fig. 6.18

Chapter 7

Energy in the Switched Two-Capacitor Circuit

In the first chapter it was mentioned that the delta current pulse $i(t) = CV_s\delta(t)$ transporting in zero time an amount of charge $Q = CV_s$, is an unbounded current. The concepts bounded and unbounded currents come from some considerations in network theory related to circuits including capacitors and switches (or, more general, stepping voltages). As such they are closely related to intriguing networks, such as the switched capacitor network of figure Fig. 7.1 that seem to violate energy conservation. They are also related to the description of energy in networks that have reduction in "free-energy" and thus to nanoelectronics.

7.1 Problem Statement

Figure Fig. 7.1 shows two capacitors with equal capacitance values. Suppose that before the switch is closed one of the capacitors is uncharged, while the other is charged with Q_0. The energy stored in the circuit before the switch is closed is

$$w_{\text{se|before}} = \frac{(Q_0)^2}{2C} + 0 = \frac{(Q_0)^2}{2C}.$$

After the switch is closed the energy stored in the circuit is

$$w_{\text{se|after}} = \frac{(Q_0/2)^2}{2C} + \frac{(Q_0/2)^2}{2C} = \frac{(Q_0)^2}{4C}.$$

We see that at $t = 0$ the charge is redistributed in the network. This suggests that there was an unbounded current at $t = 0$.

Before examining this network in detail, first the continuity property of bounded capacitor currents, the properties of capacitor voltages in networks

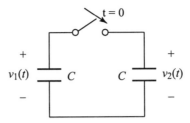

Fig. 7.1 Switched capacitor network example. Assume an initial voltage across one of the capacitors and then close the switch; the energy stored on the capacitors will be halved. Note that the voltages on both capacitors step when the switch closes.

Table 7.1 Important quantities and symbols used in this chapter

Quantity	Symbol	Unit	Symbol
time	t	second	s
charge	Q, q	coulomb	C
current	i	ampère	A
voltage difference	v	volt	V
resistance	R	ohm	Ω
capacitance	C	farad	F
energy	w	joule	J
p-operator	p		
1/p-operator	$1/p$		
impedance operator	Z		
admittance operator	Y		

with unbounded currents, and models for charged and uncharged capacitors are reviewed.

7.2 Continuity Property in Linear Networks

It is possible to predict whether a capacitor voltage $v_C(t)$ in a network (circuit) containing switches is continuous, that is wether $v(0^+) = v(0^-)$ for this capacitor. In the two capacitors network (figure Fig. 7.1), for example, we see that the capacitor voltages are discontinuous, if $v_1(0^-) \neq v_2(0^-)$. The following theorem predicts whether capacitor voltages are continuous in networks.

Theorem: [Henderson (1971)] In a linear network consisting of resistors, inductors, capacitors, switches operating simultaneously, dc voltage sources and current sources, one is sure that: the capacitor voltages are continuous except in those capacitors which, together with at least one make contact, form a loop in the network derived from the original by reducing all source intensities to zero.

7.2.1 *Continuity property of bounded capacitor currents*

For convenience, only linear time-invariant capacitors are assumed. In general, using the constitutive relation $v = C^{-1}q$ and the definition of the current $i = \mathrm{d}q/\mathrm{d}t$ the voltage across a capacitor is given by:

$$v(t) = C^{-1}q(t_0) + C^{-1} \int_{t_0}^{t} i(\tau)\mathrm{d}\tau \quad t \geq t_0. \tag{7.1}$$

More specific, if $q(0^-)$ is the charge on a capacitor with capacitance C just before closing the switch the voltage across a capacitor is:

$$v(t) = C^{-1}q(0^-) + C^{-1} \int_{0^-}^{t} i(\tau)\mathrm{d}\tau \quad t \geq 0^-. \tag{7.2}$$

Clearly, to know the voltage across the capacitor we need to know the entire past history. To examine the continuity of the capacitor voltage we focus at $t = 0$, the moment of switching.

Bounded currents can now be defined [Chua *et al.* (1987)]. If the current $i_C(t)$ in a linear time-invariant capacitor is *bounded* in a closed interval $[0^-, 0^+]$, then the voltage $v_C(t)$ across the capacitor is a *continuous* function in the open interval $(0^-, 0^+)$. In particular, for time $t = 0$ satisfying $0^- < 0 < 0^+$ holds: $v_C(0^-) = v_C(0) = v_C(0^+)$.

Proof: Substituting $t = t_0 + \Delta t$ in equation 7.1, where $t_a < t_0 < t_b$ and $t_a < t_0 + \Delta t \leq t_b$, we get:

$$v_C(t_0 + \Delta t) - v_C(t_0) = C^{-1} \int_{t_0}^{t_0+\Delta t} i_C(\tau)\mathrm{d}\tau \quad t_0 + \Delta t \geq t_0. \tag{7.3}$$

Since $i_C(t)$ is *bounded* in $[t_a, t_b]$, there is a *finite* constant M such that $|i_C(t)| < M$ for all t in $[t_a, t_b]$. It follows that the area under the curve $i_C(t)$ from t_0 to $t_0 + \Delta t$ is at most $M\Delta t$ (in absolute value), which tends to zero as $\Delta t \to 0$. Hence equation 7.3 implies that $v_C(t_0 + \Delta t) \to v_C(t_0)$ as $\Delta t \to 0$. This means that the voltage $v_C(t)$ is continuous at $t = t_0$. Filling in $t_a = 0^-$ and $t_b = 0^+$ the continuity for bounded currents at $t = 0$ is proved. (A continuity property also holds for inductor currents if the voltage across the inductor is bounded.) $\triangle$

7.3   Unbounded Currents

In this context an unbounded current is an impulsive current; an unbounded current may appear in capacitor networks with stepping ideal
voltage sources in the absence of resistive elements, such as the example in
chapter 1 (figure Fig. 1.14). In this example an initially uncharged capacitor was excited by an ideal voltage source. We found that when the voltage
source waveform is a step function stepping at $t = 0$ from 0 to V_s volt, the
current through the circuit at $t = 0$ was calculated to be

$$i(t) = CV_s\delta(t), \tag{7.4}$$

where CV_s is the charge transferred during the current pulse in this circuit.

7.3.1   *Voltages in circuits with unbounded currents*

A definition of the voltage across a capacitor *in the unbounded case*, can be
found using equation 7.2. If Δq is the charge transferred during the delta
pulse then for $t > 0$ holds:

$$v(t) = C^{-1}q(0^-) + C^{-1}\Delta q \int_{0^-}^{t} \delta(\tau)d\tau \tag{7.5}$$

$$= C^{-1}q(0^-) + C^{-1}\Delta q. \tag{7.6}$$

Δq is a positive quantity if positive charge is entering a capacitor and
negative if positive charge is leaving a capacitor.

7.4   Zero Initial Capacitor Voltage (Zero State)

Because our ultimate goal is to find a model for tunneling, including a
possible tunneling time (or transit time) the analyzes will be done in the
time domain. When the charges on all the capacitors in a circuit at $t = 0$
equal zero, that is, no initial charges are present on the capacitors, then
the capacitors in the circuit and the circuit itself can be analyzed using the
p-operator notation [1].

7.4.1   *p-Operator notation*

The p-operator and the $1/p$-operator are defined as follows:

---

[1] Originally introduced by Heaviside (1850-1925).

$$p = \frac{\mathrm{d}}{\mathrm{d}t} \tag{7.7}$$

and

$$\frac{1}{p} = \int_0^t (\ )\, \mathrm{d}\tau, \tag{7.8}$$

in which $t = 0$ is the time the evaluation starts and the initial conditions are zero. The operators operate on a function; the p and $1/p$ operators only have a meaning when applied *from the left side* of a time function $f(t)$:

$$pf(t) = \frac{\mathrm{d}}{\mathrm{d}t} f(t) = f'(t) \tag{7.9}$$

and

$$\frac{1}{p} f(t) = \int_0^t f(\tau)\, \mathrm{d}\tau. \tag{7.10}$$

p^{-1} is the inverse operator of p if $f(0) = 0$, that is, if the initial state is zero:

$$p\left(\frac{1}{p} f(t)\right) = \frac{1}{p}\left(pf(t)\right) = f(t) \tag{7.11}$$

Proof:

$$\frac{\mathrm{d}}{\mathrm{d}t} \int_0^t f(\tau)\mathrm{d}\tau = f(t).$$

This is the fundamental theorem of calculus, and holds for every f if f is continuous on $[0,t]$ and has a derivative at every point of $[0,t]$. Also

$$\int_0^t \frac{\mathrm{d}f(\tau)}{\mathrm{d}\tau}\mathrm{d}\tau = f(t) - f(0) = f(t),$$

if and only if $f(0) = 0$. (p-operators are only useful if $f(t = 0) = 0$.) $\quad \triangle$

The operators p and $1/p$ are called **differential operators** and have the same properties as the operations they stand for. Especially, they are *linear*:

$$p\left[af(t) + bg(t)\right] = apf(t) + bpg(t) \tag{7.12}$$

and

$$\frac{1}{p}\left[af(t) + bg(t)\right] = a\frac{1}{p}f(t) + b\frac{1}{p}g(t), \tag{7.13}$$

where a and b are arbitrary constants.

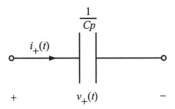

Fig. 7.2   The capacitor impedance operator; the capacitor has no initial charge.

7.4.2   *Impedance and admittance operators of the capacitor*

Now, the capacitor relationships, in terms of the operators, become:

$$i(t) = Cpv(t) \tag{7.14}$$

and

$$v(t) = \frac{1}{Cp}i(t). \tag{7.15}$$

For the capacitor we can write:

$$v(t) = Z(p)i(t), \tag{7.16}$$

where

$$Z(p) = \frac{1}{Cp} \tag{7.17}$$

is called the **impedance operator** or **operator impedance** of the capacitor. Furthermore,

$$i(t) = Y(p)v(t), \tag{7.18}$$

where

$$Y(p) = Cp \tag{7.19}$$

is called the **admittance operator** or **operator admittance** of the capacitor. The equations 7.16 and 7.18 are generalizations of Ohm's law. Using the concepts of admittance and impedance calculations on networks including *initially uncharged* capacitors can be done in a way analogous to resistive networks. Figure Fig. 7.2 shows the capacitor impedance operator; $i_+(t)$ and $v_+(t)$ are the current and voltage after $t = 0$, respectively.

7.4.3 Generalized functions

Furthermore the relations between generalized step function and generalized delta function can be expressed using the p-operators:
-the unit step function $\epsilon(t)$ defined by the relation

$$\epsilon(t) = \begin{cases} 1 & t > 0 \\ 0 & t < 0 \end{cases} \quad \text{or} \quad \epsilon(t - t_0) = \begin{cases} 1 & t - t_0 > 0 \\ 0 & t - t_0 < 0. \end{cases} \tag{7.20}$$

At $t = 0$ the unit step function is undefined but restricted between $0..1$.
- And the (Dirac) delta function or impulse function $\delta(x)$ defined as

$$\delta(t) = \begin{cases} 0 & t > 0 \\ \int_{-\infty}^{\infty} \delta(\tau)d\tau = 1 & t = 0 \\ 0 & t < 0 \end{cases} \tag{7.21}$$

define each other:

$$\int_{-\infty}^{t} \delta(\tau)d\tau = \varepsilon(t) \quad \text{or} \quad \frac{1}{p}\delta(t) = \varepsilon(t) \tag{7.22}$$

and

$$\frac{d}{dt}\varepsilon(t) = \delta(t) \quad \text{or} \quad p\varepsilon(t) = \delta(t) \tag{7.23}$$

7.5 Initial Charge Models

The vi-relation of the capacitor (equation 7.1) immediately suggests to examine a circuit in which we can switch on a current source, see figure Fig. 7.5. Before closing the switch the capacitor has a initial charge $q(0^-)$ that causes an initial voltage $v(0^-)$ across the capacitor. We are going to describe what happens if the switch is closed at $t = 0$. Therefore the switch will be modelled with the unit step function $\epsilon(t)$. Because the step function is not defined at $t = 0$ the initial voltage is the voltage just before closing the switch, $v(0^-)$. In equation 7.1 the voltage across the capacitor was written in terms of an initial value and the current waveform. Starting from here an equivalent circuit is derived including the initial value; therefore both sides of equation 7.2 are multiplied by the unit-step function $\epsilon(t)$:

$$v(t)\epsilon(t) = \left[\frac{1}{C} \int_{0^-}^{t} i(\tau)d\tau \right] \epsilon(t) + v(0^-)\epsilon(t).$$

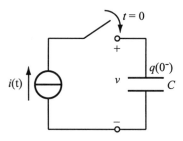

Fig. 7.3   Circuit, including a switch closing at $t = 0$, describing a charged capacitor excited by an ideal current source.

The unit-step function can be brought into the integral, thereby changing the argument of $\epsilon(t)$ to τ. This is correct because if $t < t_0$ both forms are zero, whereas if $t > t_0$ both have the value of the integral taken from 0^+ to t. Using the continuity property (and assuming bounded currents), as initial charge the known value at $t = 0^-$ can be taken.

$$v(t)\epsilon(t) = \frac{1}{C} \int_{0+}^{t} i(\tau)\epsilon(\tau)d\tau + v(0^-)\epsilon(t). \tag{7.24}$$

Now, the one-sided waveforms $v_+(t)$ and $i_+(t)$ can be defined:

$$v_+(t) \overset{\text{def}}{=} v(t)\epsilon(t) \tag{7.25}$$

and

$$i_+(t) \overset{\text{def}}{=} i(t)\epsilon(t). \tag{7.26}$$

In equation 7.24, all the charge before switching is in the second term of the expression of $v(t)\epsilon(t)$, and the first term can be identified with an uncharged capacitor at $t = 0^+$, rewriting equation 7.24 in terms of the new variables $v_+(t)$ and $i_+(t)$, and using the integration operator $1/p$, we obtain:

$$v_+(t) = \frac{1}{Cp}i_+(t) + v(0^-)\epsilon(t). \tag{7.27}$$

Looking closely at equation 7.27 we see that it is a *vi*-relationship for a capacitor consisting of two terms, two voltages. Because voltages add, these two terms can be expressed by two elements connected in series. The first term represent a capacitor, for $t > 0$, with zero initial charge, the second term represent a time-varying voltage source.

The two elements form a subcircuit shown in figure Fig. 7.4(a) and is called the series initial condition model for the capacitor. Rewriting

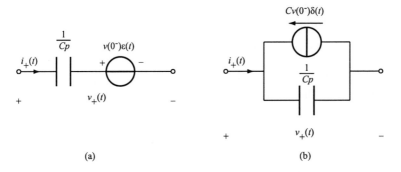

Fig. 7.4 Series (a) and parallel (impulse) models for the capacitor in the time domain. The capacitors have no initial charge.

equation 7.27 we can obtain the one-sided current $i_+(t)$ as a function of the voltage $v_+(t)$. Therefor we multiply both sides with Cp and find:

$$i_+(t) = Cp[v_+(t) - v(0^-)\epsilon(t)] = Cpv_+(t) - Cv(0^-)\delta(t), \qquad (7.28)$$

because the time derivative of the unit-step function is the delta (impulse) function. In this way we obtain the so-called parallel or impulsive initial condition model for the capacitor shown in figure Fig. 7.4. The second term $Cv(0^-)\delta(t)$ equals $q(0^-)\delta(t)$ and has the dimensions of a current. We can identify this term as a time-varying current source. In case of the series initial condition model, it is important to realize that the actual capacitor voltage is the sum of the voltages of the two elements.

Example 7.1 As an example, we consider energy relations in a circuit in which a linear capacitor is excited by an ideal dc current source:

▽ Consider a linear capacitor C with a value $C \neq 0$ excited by an ideal dc current source I_s with a constant value I_s. The result of the Kirchhoff laws and the constitutional relation $q = Cv$ is the differential equation:

$$i(t) = C\frac{dv(t)}{dt} = I_s, \qquad (7.29)$$

When the switch is closed at $t = 0$ we find for the voltage $v(t)$ for $t > 0$, $v_+(t)$:

$$v_+(t) = \frac{1}{Cp}i_+(t) + v(0^-)\epsilon(t) \qquad (7.30)$$

filling in the value for $i_+(t)$ and using the continuity property of the capacitor, and $q(0^-) = q(0^+)$, for bounded currents:

$$v_+(t) = \frac{I_s}{C}t + v(0^+)\epsilon(t) = \frac{I_s}{C}t + \frac{q(0^+)}{C}\epsilon(t), \qquad (7.31)$$

in which the initial charge $q(0^+)$ just after the switch is closed at $t = 0$ is written down explicitly. For zero initial charge we find:

$$v_+(t) = \frac{I_s}{C}t \quad \text{or} \quad v(t) = \frac{I_s}{C}t, \ t > 0. \tag{7.32}$$

Again, we consider the energy relations in the circuit, including a initial charge $q(0^+)$ at $t = 0^+$. The increase of the amount of energy stored by the capacitor w_{se} is:

$$\Delta w_{se} = \frac{q(t)^2}{2C} - \frac{q(0^+)^2}{2C} = \frac{(q(0^+) + \Delta q(t))^2}{2C} - \frac{q(0^+)^2}{2C} \tag{7.33}$$
$$= \frac{q(0^+)\Delta q(t)}{C} + \frac{(\Delta q(t)^2)}{2C}.$$

Using $v(0^+) = q(0^+)/C$ and $\Delta q(t) = I_s t$ we obtain:

$$\Delta w_{se} = v(0^+)I_s t + \frac{I_s^2 t^2}{2C}. \tag{7.34}$$

At the same time, the energy delivered by the source is:

$$w_s = I_s \int_{0+}^{t} v_+(\tau)d\tau. \tag{7.35}$$

Looking to figure Fig. 7.5 we can write:

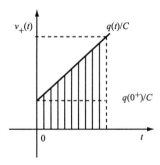

Fig. 7.5   The voltage across the capacitor, and thus the voltage across the current source, as a function of time.

$$w_s = I_s \frac{q(0^+)}{C}t + I_s \frac{1}{2}\frac{\Delta q(t)}{C}t = v(0^+)I_s t + \frac{I_s^2 t^2}{2C}, \tag{7.36}$$

we see that the extra amount of energy stored on the capacitor was delivered by the source.

$\triangle$

7.6 Solution A: Bounded Currents

For the energy problem of the capacitor network of figure Fig. 7.1 there exists two solutions, the first considers only bounded currents, the second only unbounded currents. First, we have the standard textbook solution in which a resistor is placed between the two capacitors. By inserting the resistor, the current in the circuit becomes bounded, the voltages across the capacitors will only change gradually, continuity of the capacitor currents is guaranteed, and the resistor absorbs the missing energy.

Again, assuming $v_2(0^-) = 0$ V we can find the bounded current in the circuit (remember that, in this example, both capacitances are the same and equal to C) by using the series initial charge model for the charged capacitor:

$$i(t) = \frac{v_1(0^-)}{R} \exp\left(\frac{-2t}{RC}\right) \qquad t > 0 \qquad (7.37)$$

and for the energy dissipated in the resistor $w_R = R \int_0^t i^2(\tau)d\tau$:

$$w_R(t) = -\frac{C[v_1(0^-)]^2}{4} \left[\exp\left(\frac{-4t}{RC}\right) - 1\right] \qquad t > 0. \qquad (7.38)$$

We see that in steady-state the amount of energy dissipated in the resistor is $Cv_1(0^-)^2/4$, which is equal to the amount of energy stored on both capacitors *after* the switch is closed and half of the energy stored on capacitor 1 *before* the switch was closed. We see that energy is conserved at all times.

7.7 Solution B: Unbounded Currents

Second, there exists a solution in which the current is unbounded. This solution starts by representing both capacitors by their equivalent initial condition models, in which the charged capacitor is modeled as an uncharged capacitor in series with a voltage source representing the charge on the capacitor. Because the change of charge on the capacitors will now be modeled as a step function in the initial conditions (from $v(0^-)$ to $v(0^+)$) the resulting current will be an impulse function—the current through the capacitors being proportional to the derivative of the voltage across them. For example, Davis [Davis (1994)] gives an excellent treatment.

For our circuit with equal capacitances we find that the initial charges are distributed over both capacitances in such a way that the Kirchhoff voltage law holds: because the capacitances are the same the voltages across both capacitors must be the same. Consequently, the charge on both capacitors will be

$$\frac{q_1(0^-) + q_2(0^-)}{2}.$$

The delta current pulse, thus, moved an amount charge of

$$q_1(0^-) - \frac{q_1(0^-) + q_2(0^-)}{2} = \frac{q_1(0^-) - q_2(0^-)}{2},$$

and leads directly to the current

$$i(t) = \frac{1}{2}[q_1(0^-) - q_2(0^-)]\delta(t). \tag{7.39}$$

or

$$i(t) = \frac{1}{2}C[v_1(0^-) - v_2(0^-)]\delta(t). \tag{7.40}$$

This current is the current redistributing the charge at $t = 0$. Again we can consider the case that the initial charge on the second capacitor is zero. We see from equation 7.40 that half of the initial charge on the first capacitor is transported to the second with an impulsive current $i(t) = (1/2)Cv_1(0^-)\delta(t)$. Now, just as in the first solution we are going to consider the energy in this circuit: what is absorbing half of the energy in the absence of a resistor?

7.7.1   *Energy generation and absorption in circuits with un-bounded currents*

The capacitor network in chapter 1, subsection 1.4.3, showed the energy generation and the energy balance in the simplest network involving an impulsive current, namely, a single capacitor charged by an ideal voltage source. The network charges an initially uncharged capacitor with a charge q. The ideal voltage source, therefore, steps from 0 V to $C^{-1}q$ at $t = 0$. An impulsive current will transfer an amount of q Coulomb towards the capacitor—and, of course, at the same time carry off an amount of $(-q)$ from the negative side of the capacitor plates. The energy stored the capacitor is raised from 0 J to $q^2/(2C)$. We calculated the energy delivered by the source w_s, using equation 1.33:

$$w_s = \int vi \, dt = C^{-1}q^2 \int \varepsilon(t)\delta(t)dt = \frac{q^2}{2C}.$$

We noticed that, of course, energy stored on the capacitor is provided exactly by the source; energy is conserved in this circuit.

Note that we are now able to calculate the energy generated by a stepping voltage source at $t = 0$ through which a delta current pulse flows at $t = 0$. If the step had been down the source would have been absorbing energy; this should not be a surprise, because the source is just a non-linear resistor being able to be both passive or active.

The result can be generalized by considering a source stepping from v_{before} to v_{after} through which a delta current pulse $i(t) = q\delta(t)$ is flowing. The stepping source can be represented by:

$$v(t) = v_{\text{before}}(t) + (v_{\text{after}}(t) - v_{\text{before}}(t))\varepsilon(t) \tag{7.41}$$

So, the energy delivered by the source in this case is

$$w_s = \int vi\,dt = q \int \left[v_{\text{before}}(t) + (v_{\text{after}}(t) - v_{\text{before}}(t))\varepsilon(t)\right]\delta(t)dt \tag{7.42}$$

using that all the voltages are constant just before and after the event leads to

$$w_s = qv_{\text{before}} + q(v_{\text{after}} - v_{\text{before}}) \int \varepsilon(t)\delta(t)dt = qv_{\text{before}} + \frac{q}{2}(v_{\text{after}} - v_{\text{before}}),$$

giving the important result

$$w_s = \frac{q}{2}(v_{\text{before}} + v_{\text{after}}). \tag{7.43}$$

7.7.2 *Energy conservation*

Now, we can attack the energy problem in case we have an unbounded current. The energy balance consists of the energy storage before the switch is closed (that is, the energy at $t = 0^-$), the energy storage after the switch is closed (that is, the energy at $t = 0^+$), and the energy generated and/or dissipated *during* closing of the switch at $t = 0$. Again we consider for simplicity the charge on the first capacitor to be q and the charge on second capacitor to be zero.

We find: the energy stored before closing is $q_1^2/(2C)$; the energy stored after closing is $2(q_1/2)^2/(2C) = q_1^2/(4C)$. For the energy generated and absorbed at $t = 0$ we realize that during closing a current $i(t) = (q_1/2)\delta(t)$ flew through the circuit while the voltage sources representing the initial conditions—in this case, the values at $t = 0^-$ and $t = 0^+$—were both stepping: the source of the first capacitor stepped down from $q_1 C^{-1}$ to $(q_1/2)C^{-1}$, and the source of the second capacitor stepped up from 0 to $(q_1/2)C^{-1}$. Using Equation 7.43 we calculate that the source of the first capacitor while stepping down *absorbed* an amount of energy equal to

$$\frac{q_1/2}{2}\left(\frac{q_1}{C} + \frac{q_1}{2C}\right) = (3/8)(q_1^2/C),$$

while the source of the second capacitor while stepping up *generated* an amount of energy equal to

$$\frac{q_1/2}{2}\left(0 + \frac{q_1}{2C}\right) = (1/8)(q_1^2/C).$$

Consequently, the net result is that during switching the sources absorbed

$$\frac{q_1^2}{C}\left(\frac{3}{8} - \frac{1}{8}\right) = q_1^2/(4C),$$

that is, together they model the absorbtion of the missing energy at $t = 0$!

7.8 Unbounded or Bounded Currents Through Circuits

We have seen that the energy delivered by a stepping voltage source for charging a capacitor with an unbounded current is:

$$w_s = \frac{q^2}{2C} \qquad \text{unbounded current.} \qquad (7.44)$$

This is *fundamentally* different from the case a bounded current is used. To obtain a bounded current in this circuit a resistor must be added, in series with the capacitor, to limit the current. The energy delivered by a stepping voltage source in the latter case is

$$w_s(t) = \int i(t)C^{-1}q\varepsilon(t)dt = C^{-1}q\int i(t)dt \qquad t > 0. \qquad (7.45)$$

In steady state $\int i(t)dt$ will be the total amount of transported charge q, so that we obtain in steady state:

$$w_s = \frac{q^2}{C} \qquad \text{bounded current.} \qquad (7.46)$$

The resistor will dissipate the difference between the amount of energy delivered by the source and the amount of energy stored by the capacitor.

In fact the above described difference *proofs* that an unbounded current cannot flow (other than strictly local—we will find this in the description of the tunnel current in the next chapter) through a circuit path in which resistors are present. The proof is based on energy arguments.

Problems and Exercises

7.1 What is the amount of energy necessary to charge a capacitor (a) with an ideal current source? (b) With an ideal voltage source. (c) With a Thévenin equivalent. (d) With a Norton equivalent.

7.2 Consider the network of figure Fig. 7.6. At $t = 0$ the island charge changes from $0 \rightarrow$ e. Take $C = 1$. Calculate the change in energy at both capacitors and calculate the energy provided by the source at $t = 0$. (Ans. 0, $e^2/4$, $-e^2/4$).

(This problem proves that for this circuit energy is conserved. Later on we will see that this circuit actually mimics tunneling of a single-electron from the island through the upper capacitor.)

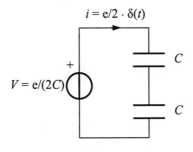

Fig. 7.6

7.3 Consider a switched two capacitor circuit (with $C_1 = C_2 = C$) without a resistor and with a resistor. (a) Prove that in the circuit without a resistor the Kirchhoff voltage law is violated for any solution in which the charge is not equally shared by both capacitors after the switch is closed. (b) Prove that intermediate charge distributions are possible, without violating the KVL, if a resistor is included.

Chapter 8

Impulse Circuit Model for Single-Electron Tunneling—Zero Tunneling Time

The fundamental physical principle underlying circuits including single-electron tunneling junctions (called SET circuits) is the Coulomb blockade, resulting from the quantization of charge, and the associated phenomenon of Coulomb oscillations. The basic physical phenomenon under consideration is the quantum mechanical tunneling of electrons through a small insulating gap between two metal leads. A single tunnel event can best be understood as an electron sitting at *both* sides of the barrier *at the same time*, but the probability of finding the electron going from one side to the other is decreasing on one side of the barrier and increasing on the other side during tunneling. As a consequence of such a description charge is moved inside the tunneling junction.

If present, all metal islands will be considered large enough not to show energy quantization. The metal-insulator-metal structure through which the electrons tunnel is called a single-electron tunneling (SET) junction. This tunneling is considered to be stochastic, that is, successive tunneling events across a tunnel junction are uncorrelated, and is described by a Poisson process. Unless stated explicitly, tunneling through a potential barrier is considered to be non-dissipative (the tunneling process through the barrier is considered to be elastic).

Having described the physical and circuit theoretical background in the previous chapters, in the coming three chapters a circuit theory for metallic single-electron tunneling junctions is described [Hoekstra (2004, 2007b,a)]. The starting points are the experimental results presented in chapter 2. If *many* electrons are tunneling it is possible to describe this tunneling with a tunnel current, as was the case in the description of tunneling in a tunnel diode or in the description of a metal-insulator-metal diode, when both the capacitance of the plates and the applied bias voltage are sufficiently large [1]. In a circuit theory those devices can then be modeled as (nonlinear) resistors. The new phenomena Coulomb blockade and Coulomb oscillations, however, are the result of tunneling of *single* electrons.

In this chapter a circuit model is developed for a (single) tunneling electron with a zero tunneling time and without considering the probabilistic nature of the tunneling process. Beside this, considered are only the ideal energy sources, without internal losses: ideal current source and ideal voltage source. The result of the modeling will be one single circuit description of the tunneling electron, but two descriptions, however, of the associated tunnel current and the associated Coulomb blockade. The appearance of the two descriptions is a consequence of the appearance of a bounded current in the case of an ideal current source but a unbounded current in case of an ideal voltage source.

The analysis of tunnel junctions excited by real (nonideal) sources opens the way for one single description, because all associated currents will be bounded. In chapter 9 this analysis is done. The uniform treatment will also give the possibility to derive an expression for the Coulomb blockade for a nonzero tunneling time. Still, only a single tunneling electron is considered without stochastic behavior. The inclusion of time in the analysis leads to the description of a tunnel current caused by tunneling of many electrons in the same time-slot, although the first electron to tunnel will determine the value of the Coulomb blockade. The tunneling of many electrons lead to resistive behavior, and modeling of the tunnel junction as a resistor for voltages with absolute values above the Coulomb blockade. In chapter 10 we return to the description of a single tunneling electron to derive expressions for the Coulomb blockade in multi-junction circuits.

In this chapter, an equivalent circuit element representing the tunneling of a single electron is derived and possible physical justifications are presented. And based on the conservation of energy in the circuits—a fundamental circuit theorem—the condition for tunneling is derived. The resulting model represents the tunnel event of a single tunneling electron by an impulse current source, the junction by a charged capacitor, and the tunnel condition based on local circuit parameters.

---

[1]A metal-insulator-metal diode with small capacitances and small voltages can show Coulomb blockade (in case of bounded currents), as will be discussed further on.

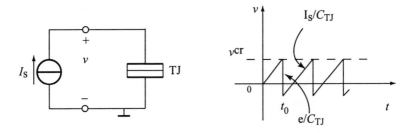

Fig. 8.1 SET junction excited by an ideal DC current source, switched on at $t = 0$, showing Coulomb oscillations; tunneling time is zero.

8.1 SET Junction Excited by an Ideal Current Source— Zero Tunneling Time

As presented in subsection 2.2.4, a (or an array of) tunneling junction(s) shows Coulomb oscillations when excited by a current source, that is, the voltage across the tunneling junction oscillates. To explain this behavior it is generally accepted that the SET junction acts as a capacitor until a critical voltage v_{TJ}^{cr} across the tunnel junction is reached. This junction capacitance is charged by the current source. Upon reaching the critical voltage an electron tunnels through the junction, thereby lowering the voltage across the junction by e/C—where e is the elementary electron charge.

8.1.1 Coulomb oscillations

Figure Fig. 8.1 shows the circuit and its dynamic behavior. Using the constitutional relationship of the capacitor $v = C^{-1}q$ and the definition of the current $i = dq/dt$, for a constant current source with value I_s we can immediately calculate the oscillation frequency:

$$f = \frac{1}{\Delta t} = \frac{C^{-1}I_s}{\Delta v} = \frac{C^{-1}I_s}{C^{-1}e} = \frac{I_s}{e}. \tag{8.1}$$

8.1.2 Tunneling of a single electron modeled by an impulsive current

By modeling the tunnel(ing) junction with a linear (tunneling) capacitor we require that the equivalent circuit will always be described by linear circuit theory. During charging, the equivalent circuit is that of a charging capacitor; and all energy stored on or retrieved from the capacitor is pro-

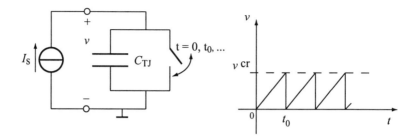

Fig. 8.2 A first switched circuit to describe the current excited SET junction and its response.

vided or absorbed by the current source, respectively. To find an equivalent circuit for the tunnel(ing) junction we first construct a linear circuit that effectively models the sawtooth behavior. Because we know that before tunneling the tunnel junction can be modeled by a linear capacitor—as a first try—we come up with the parallel combination of a capacitor and a switch, which closes for a very short period (approaching zero seconds) at any time the voltage v reaches v^{cr}. Figure Fig. 8.2 shows this circuit and its response. Upon closing the switch any charge on the capacitor will be drained immediately; using the theorem on the continuity property in linear network of section 7.2 we know that the voltage on the capacitor will be discontinuous.

It is immediately clear that this circuit could never produce the negative voltages needed for the correct SET behavior of figure Fig. 8.1 without changing the direction of the current source. We can, however, obtain a qualitative correct equivalent circuit, by adding a voltage source with value

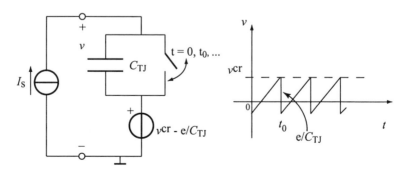

Fig. 8.3 A second switched circuit to describe the current excited SET junction and its response.

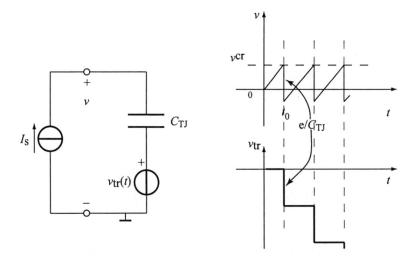

Fig. 8.4 A third circuit to describe the current excited SET junction and its response.

$v^{\mathrm{cr}} - e/C_{TJ}$. Figure Fig. 8.3 shows this second switched equivalent circuit for the SET junction excited by an ideal current source, and its response.

To remove the switch in the circuit we combine the capacitor with a variable voltage source in the way as shown in figure Fig. 8.4. The voltage source has a staircase-like behavior as depicted in the figure. This circuit is equivalent to the previous one. This can be understood by realizing that the resulting voltage across the capacitor *and* additional source is the superposition of the voltage from the continuing charging of the capacitor and the stepping source $v_{\mathrm{tr}}(t)$.

To make a closer link to the actual electron tunneling process we first consider the behavior of the circuit at $t = t_0$ only, that is we are going to consider the behavior of the equivalent circuit for $0 \leq t \leq \frac{3}{2} t_0$. We observe that within this time span the added voltage source $v_{\mathrm{tr}}(t)$ steps from 0 V down to $-e/C_{\mathrm{TJ}}$ V. Since stepping of the added voltage source—that mimics a tunnel event—takes place at $t = t_0$, we want to describe what happens during tunneling, that is, from $t = t_{0-}$ till $t = t_{0+}$.

To simplify the notation in the following derivations the subscript TJ is left out of the expression for the tunnel junction capacitance, and the capacitor is described in the time-domain using the (Heaviside) p-operator.

Using the general current-voltage relationship for a capacitor C, taking into account that we started to charge the capacitor at $t = 0$, we have:

$$v(t) = \frac{1}{C_p} i(t) + v_{tr}(t) = \frac{1}{C} \int_0^t i(\alpha) d\alpha - \frac{e}{C} \varepsilon(t - t_0) \tag{8.2}$$

that is, within our time span: if $t < t_{0-}$:

$$v(t) = \frac{1}{C} \int_0^t i(\alpha) d\alpha, \tag{8.3}$$

and for $t > t_{0+}$:

$$v(t) = \frac{1}{C} \int_0^{t_{0-}} i(\alpha) d\alpha + \frac{1}{C} \int_{t_{0-}}^{t_{0+}} i(\alpha) d\alpha + \frac{1}{C} \int_{t_{0+}}^t i(\alpha) d\alpha - \frac{e}{C}. \tag{8.4}$$

In equation 8.4 the first integral represents, of course, the voltage just before $t = t_0$ as a consequence of the net amount of charge stored on the capacitor during the interval $<0, t_{0-}>$. We denote this voltage as $v(t_{0-})$. The second term indicates the additional change in the voltage across the capacitance as a consequence of the step action. This term is considered to be zero, because due to the conservation of charge on the capacitor during the step action, the absence of switching impulses to the capacitor is assumed. The third term represents the charging of the capacitor during the interval $< t_{0+}, t >$, and the last term is the decrease in voltage after the step.

Now, both equations are combined to obtain:

$$v(t) = v(t_{0-}) + \frac{1}{C} \int_{t_{0+}}^t i(\alpha) d\alpha - \frac{e}{C}, \tag{8.5}$$

To describe the voltage after the tunnel event again, we multiply both sides of equation 8.5 by the unit step function $\varepsilon(t - t_0)$, and get:

$$v(t)\varepsilon(t - t_0) = v(t_{0-})\varepsilon(t - t_0) + \left[\frac{1}{C} \int_{t_{0+}}^t i(\alpha) d\alpha \right] \varepsilon(t - t_0) - \frac{e}{C} \varepsilon(t - t_0) \tag{8.6}$$

The lower limit of the integral is adapted to describe an uncharged capacitor by again considering, first, that for $t < t_0$ the step function is zero, and, second, that at $t = t_0$ the contribution to the voltage change due to the capacitor is zero. This last fact also results in a zero contribution, because at $t = t_0$ the value of the unit step function is undefined but restricted between 0 and 1. We obtain:

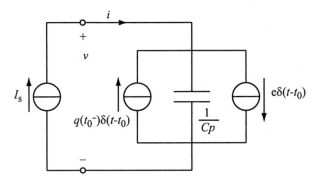

Fig. 8.5   Parallel equivalent circuit in the time domain, using impulsive current sources.

$$v(t)\varepsilon(t - t_0) = v(t_{0-})\varepsilon(t - t_0) + \frac{1}{C}\int_0^t i(\alpha)\varepsilon(\alpha - t_0)d\alpha - \frac{e}{C}\varepsilon(t - t_0), \quad (8.7)$$

writing $v_+(t) = v(t)\varepsilon(t - t_0)$ and $i_+(t) = i(t)\varepsilon(t - t_0)$, we find:

$$v_+(t) = v(t_{0-})\varepsilon(t - t_0) + \frac{1}{Cp}i_+(t) - \frac{e}{C}\varepsilon(t - t_0). \quad (8.8)$$

We see that, for $t > t_0$, the iv-relationship consists of three terms: the actual charged capacitor voltage is the sum of a uncharged capacitor at $t > 0$, an independent voltage source, and the tunnel event is modelled by a stepping source. This equivalent circuit is referred to as a *series tunneling model*, because its elements are series connected. A parallel model can be obtained:

$$i_+(t) = Cp\left[-v(t_{0-})\varepsilon(t - t_0) + v_+(t) + \frac{e}{C}\varepsilon(t - t_0)\right]$$
$$= -Cv(t_{0-})\delta(t - t_0) + Cpv_+(t) + e\delta(t - t_0), \quad (8.9)$$

a so-called *parallel tunneling model*. We notice that as a consequence of the presence of the stepping voltage source in the series tunneling model an impulsive current source appears in the parallel model with a value of $e\delta(t - t_0)$. If we use that $q = C_{TJ}v$, figure Fig. 8.5 shows the parallel equivalent circuit reflecting equation 8.9.

8.1.3   *Energy is conserved: critical voltage*

If we consider the delta current source representing the charge in the capacitor at $t = t_{0-}$ and the uncharged capacitor as a single charged capacitor

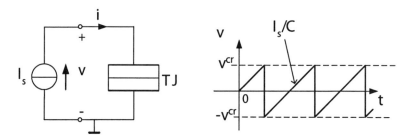

Fig. 8.6 If energy is conserved: SET junction excited by an ideal DC current source, switched on at $t = 0$, showing Coulomb oscillations; tunneling time is zero.

at $t = t_{0-}$, the parallel equivalent model reduces to a charged capacitor in series with a delta current source, representing the tunneling of an electron in the time slot $[t_{0-}, t_{0+}]$. Using this model we can easily calculate the amount of energy involved in the tunneling process when modeled in this way. From subsection 7.7.1 we recall that when the voltage steps when the current through the equivalent circuit is a delta pulse, the energy involved can be expressed as:

$$w_{\mathrm{TJ}} = \frac{e}{2}[v(t_{0-}) + v(t_{0+})]. \tag{8.10}$$

We can use this equation to find t_0. To keep the circuit model within the domain of circuit theory, we require that Kirchhoff's laws have to hold and thus Tellegen's theorem applies to the circuit at $t = t_0$; that is, energy must be conserved in the circuit during tunneling. During tunneling the current source doesn't supply any energy to the circuit, because the tunneling time is considered to be zero. Consequently the *only* solution to bring the equivalent circuit within the circuit domain is to require that w_{TJ} is also zero. Using equation 8.10 this immediately leads to the requirement that $[v(t_{0-}) + v(t_{0+})] = 0$:

$$v(t_{0+}) = -v(t_{0-}) \quad \text{tunnel condition in case of a current source.} \tag{8.11}$$

Figure Fig. 8.6 shows the circuit and its behavior. The predicted behavior matches the experiments on the current source driven SET junction described in chapter 2. The circuit shows Coulomb oscillations with a frequency $f = I_s/e$. This tunnel condition together with the requirement that a whole electron charge has to be transferred during the tunnel event leads to the result that the critical voltage of a tunnel junction excited by an ideal current source in case of a zero tunneling time equals:

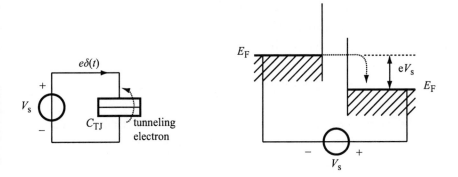

Fig. 8.7   Single-electron tunneling junction excited by an ideal voltage source above the critical voltage and energy-band diagram showing the tunneling of a single electron from the Fermi level.

$$v_{\mathrm{TJ}}^{\mathrm{cr}} = \frac{e}{2C_{\mathrm{TJ}}}, \quad \text{ideal current source;} \quad \tau = 0. \tag{8.12}$$

8.2   SET Junction Excited by an Ideal Voltage Source

Second, we consider a metal-insulator-metal tunnel junction excited by an ideal constant voltage source. For large capacitances the response is well-known and shows the vi-characteristic of a linear resistor at low source voltages. We assume the tunneling electrons to tunnel through the tunnel junction one-by-one. Figure Fig. 8.7 shows the circuit and the energy-band diagram. We follow a single electron that tunnels from the Fermi level at the negative side and investigate whether we can use the impulse circuit model for tunneling of this single electron. Due to the fact that our model circuit only consists of an ideal voltage source and a capacitor representing the junction's capacitance, but that resistors are absent, the unbounded impulsive tunnel current will propagate through the whole circuit. Analyzing this circuit we find:

$$\text{KVL:} \quad v_{\mathrm{TJ}} = V_{\mathrm{s}} \tag{8.13}$$

$$\text{KCL:} \quad i(t) = e\delta(t) \tag{8.14}$$

Also the energy balance in the model circuit is well described. It is clear from the energy-band diagram that the tunneling electron, in this case, dissipates an amount of energy equal to eV_{s} during the transition from the

Fermi level at the negative side of the source to the Fermi level at the positive side of the source; the electron is a hot electron.

The energy provided by the source is

$$\int V_\mathrm{s} e\delta(t)dt = eV_\mathrm{s} \int \delta(t)dt = eV_\mathrm{s} \tag{8.15}$$

So, the source provides the dissipated energy at the time of tunneling. What remains, now, is to prove that the equivalent circuit for the tunneling junction also absorbs an amount of energy of eV_s.

Again we consider the equivalent circuit, representing tunneling, to be a capacitor in parallel with a impulsive current source $i(t) = e\delta(t)$ and find the energy dissipated by this equivalent circuit

$$w_\mathrm{TJ} = \frac{e}{2}\left[v(0^-) + v(0^+)\right] = eV_\mathrm{s},$$

because $v(0^-) = v(0^+) = V_\mathrm{s}$. Energy is again conserved.

8.2.1   *Critical voltage*

It is clear from the above reasoning that for the single tunnel junction excited by an ideal voltage source the critical voltage must be zero. The current through a tunneling junction as a consequence of the tunneling of only one electron in zero time is the unbounded current $i(t) = e\delta(t)$. For any nonzero constant voltage source V_s the amount of energy provided by the source equals eV_s and the energy absorbed at the tunnel junction is also eV_s :

$$v_\mathrm{TJ}^\mathrm{cr} = 0, \quad \text{ideal voltage source;} \quad . \tag{8.16}$$

8.3   Basic Assumptions

In this section the basic assumptions of the impulse circuit modelare presented and their physical interpretations are suggested. They form the starting point for further elaboration of the model for tunnel junctions circuits including a tunneling time and for multi-junction circuits; both treated in following chapters.

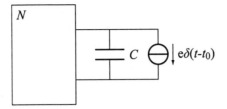

Fig. 8.8   Equivalent cirucit for the single-electron tunneling junction.

8.3.1   *During tunneling the equivalent circuit will have an impulsive component*

A first idea for the equivalent circuit for the metallic tunneling junction is to describe the change in the charge on the junction capacitance, due to tunneling, as a step function and subsequently to model the change in charge, that is, the local tunnel current for a single electron, as a (Dirac) delta function (or unit impulse function):

$$i_{\text{TJ}}(t) = e\delta(t - t_0). \tag{8.17}$$

In this way, the tunneling charge is quantized naturally by introducing the elementary charge e. The total charge that tunneled is now:

$$q(t_0) = e \int_{-\infty}^{\infty} \delta(\tau - t_0)\mathrm{d}\tau = e. \tag{8.18}$$

The delta function guarantees that $e\delta(t - t_0)$ is the current that belongs to the transfer of exactly a single electron at $t = t_0$ from one side of the junction to the other. One can easily check that the dimension of $e\delta(t-t_0)$ is a current. Remember the definition of the delta function; it shows that the dimension of the delta function is frequency, in our case. Alternatively, the parallel delta impulse current can be replaced by a serial stepping voltage which stands for the stepping initial charge on the junction during the tunnel event.

Only in case of circuits consisting of tunnel junctions, capacitors, and ideal voltage sources an impulsive, unbounded, current due to the tunneling of a single electron can propagate through the whole circuit. In all other cases in the impulsive current cannot exist outside the equivalent circuit; in the remainder of the circuit, N, the current must then be bounded. Figure Fig. 8.8 shows the equivalent circuit.

The complete equivalent circuit also contains a capacitor to model the metallic SET junction in Coulomb blockade. The formulation of the tunnel current as a delta function also allows us to incorporate the tunneled charge (or a fraction of the tunneled charge) as initial charge on the capacitor in this way making it possible to use a circuit theory that includes the tunnel event [2].

8.3.2 *In SET circuits energy is conserved*

The circuits including the single-electron tunneling junctions are considered to be lumped circuits; the geometry, shape, and physical extent of the elements are irrelevant. For the "standard" elements: resistors, capacitors, inductors, and transformers this means that any interconnection of them is such that the dimensions of the circuits are small compared with the wavelength associated with the highest frequency of interest. For the tunneling elements this requires that we have to find an equivalent circuit composed of "standard" elements.

A circuit theory for these circuits is based on the Kirchhoff laws and Tellegen's theorem. The current law (KCL) ensures the conservation of charge in the circuit; the voltage law (KVL) ensures that besides the conservation of charge energy is conserved too. The combined validity of the two Kirchhoff laws in Tellegens theorem ensures the conservation of energy (or power) in any circuit. As a theorem based on the topology of circuits, Tellegens theorem is general: it applies to all (sub)circuits and elements, whether they are linear or nonlinear, time-invariant or time variant, passive or active. The excitation might be arbitrary; the initial condition are also arbitrary. If any circuit theorem is capable of including the new quantum devices, it must be Tellegen's theorem. The conservation of energy implies that all the energy delivered by the sources in the circuit is dissipated or stored in the other circuit elements. Reduction of energy in the circuit is not possible.

8.3.3 *Tunneling through the barrier is nondissipative*

Although tunnelling through the barrier can be dissipative due to scattering at impurities in the insulator material, this is a second order effect. In the impulse model tunneling through the barrier is considered to be non-

---

[2]In circuit theory, chapter 7, initial charge on a capacitor can be represented *in the time domain* by a series step voltage source or by a parallel impulsive current source.

dissipative (elastic), like band-to-band tunneling in a tunnel diode. The tunneling electrons have the same definite energy just before, during, and immediately after the penetration of the barrier.

8.3.4   *Tunneling is possible when a reservoir filled with electrons is facing empty energy levels*

Experiments measuring tunnel currents, like the experiments on a tunnel diode or a metal-insulator-metal diode, show that if the tunneling junction is excited by a voltage source the *vi*-characteristic is linear for small applied voltages and relatively large junction capacitances. This behavior can be explained successfully by a model in which the tunneling electrons on one side of the junction are facing empty energy states at the other side. In general in describing a tunnel *current* between two metals, there are a large number of states near a given energy level from which an electron can tunnel but we can never determine the actual pair of initial and final states. In these devices we cannot specify *a priori* which electron will tunnel, or *a posteriori* which electron has tunneled [Thornber *et al.* (1967)]. The solution of such problems involves the application of Fermi's golden rule, which gives the probability per unit time for electron transit. Assuming the difference in Fermi-energies fixed, it predicts the average current to be proportional to the applied voltage [Ingold and Nazarov (1992)]. For explaining the existence of a Coulomb blockade or Coulomb oscillations, however, we need to consider single electrons, because it is the possibility or impossibility of individual electrons to tunnel that causes these phenomena. Beside this, considering the junction as a capacitor (junction in Coulomb blockade) gives rise to dynamically changing Fermi-levels. In the impulse circuit model, as a guiding principle is taken that single-electron tunneling is possible if the electron is facing an empty energy level during the complete transition.

8.3.5   *In metals conduction electrons move on the Fermi-level*

The quantum mechanical description of the conduction considers the wave nature of the electron, it involves the scattering of electron waves by the lattice. Quantum mechanical calculations show that, for a *perfectly* ordered crystal, there is no scattering of electron waves. Therefore scattering occurs due to deviations from a perfectly ordered crystal. The most important deviations being thermal vibrations of the lattice ions or impurities in the metal. Another important result is that *only electrons at the very vicinity of the Fermi level* can contribute to the current density.

In chapter 5 we saw that quantum mechanically the electrical conductivity in metals is explicitly written as:

$$\sigma = \frac{e^2}{3} \Lambda_F v_F N(\epsilon_F),$$

Where Λ_F is the mean free path of the conduction electron at the Fermi level and is equal to $\Lambda_F = \tau v_F$, τ is the average collision time. The equation shows that the conductivity is determined only by electrons at the Fermi-level and is proportional to the number of electrons at the Fermi level $N(\epsilon_F)$, their velocity v_F, and the mean free path Λ_F.

The equation holds for an electric field applied to a metal at a constant temperature (as a part of the linearized equation) and considers scattering of electron waves for which only deviations from a perfectly ordered lattice are important (the so-called relaxation time approximation). The equation, although, predicts that the conductance is almost independent of the temperature and the scattering of the conduction electron is considered to be elastic. Scattering with a static source of disturbances like impurity atoms or dislocations can be treated as being elastic, but scattering with lattice vibrations occurs through the exchange of energy with phonons. Detailed calculations of this latter mechanism show that these contributions give rise to the correct temperature dependence of the conductance at finite temperatures.

So far the general assumptions that underlie the impulse model. In the next section the tunnel condition that links to the basic assumptions is presented.

8.4   Conditions for Tunneling

Now, consider some experiments. First, we see that the *iv*-characteristics show a resistive behavior for junctions excited by a (real) voltage source with a value above a critical voltage. A well studied example is the single-electron tunnelling transistor, consisting of two tunnel junctions and a (gate) capacitor that controls the charge distributions on the island formed by both the two junctions in Coulomb blockade and the capacitor. Second, in other experiments the tunnel junctions behave as (linear) capacitors, for example, as is seen in experiments on Coulomb oscillations. We look into

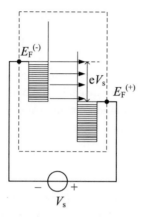

Fig. 8.9 Energy diagram for tunneling between two metals with fixed Fermi levels, that is, charging effects due to electron tunneling can be neglected. Tunneling is possible only when filled energy states are facing empty energy states. The electrical behavior is resistive; for small source voltages, the tunneling junction can be modeled as a linear resistor.

both cases.

8.4.1 *Resistive behavior*

For a circuit theoretical description we consider the entire junction device, that is, the junction itself and a part of the metals, as a black box. To describe electrically this black box we assume that: across the entire junction device (seen as a black box), electrons move from Fermi level to Fermi level. In this way the source provides the energy eV_s to bring an electron from the Fermi level on the positive side to the Fermi level on the negative side, see figure Fig. 8.9. The motivation for this assumption is that the conduction electrons move on the Fermi levels in the metallic interconnect, and that, by definition the Fermi levels before and after the tunneling event are the same if and only if the behavior is resistive. For in this case, at any point in time the current through the junction (or resistor) is determined by the potential across it only at the same point in time. Any dynamics can only be described by adding dynamic elements, such as (parasitic) capacitors or (parasitic) inductors.

8.4.2   *Hot-electron model at the positive side*

At any non-zero value of the voltage source, the electrons at the negative side are facing empty energy states at the positive side and will have a probability to tunnel. For the time being we only consider $T \approx 0K$; the influence of the temperature can be easily included using Fermi-Dirac statistics. First we follow the tunneling of conduction electrons, that is, electrons arriving at the tunneling junction at the Fermi level on the negative side (the uppermost arrow in figure Fig. 8.9). After tunneling these electrons enter the metal at the positive side well above the Fermi level. This is, of course, because we consider nondissipative tunneling. Electrons appearing after tunneling above the Fermi level possess a high kinetic energy, and are usually called hot electrons. In our case the conduction electrons entering the metal at the positive side are "hotter" by just about eV_s. The electrons travel into the metal thereby losing their excess energy. The two types of collision responsible for this are electron-phonon (interaction with the lattice) and electron-electron collisions. The latter is important in the metal because of the high free-electron concentration. Repeated collisions drop the energies of all hot electrons, and eventually, some distance away from the interface, they become part of the normal free-electron distribution. Within the normal free-electron distribution the conduction electrons can be found at the Fermi level again. The hot electron concentration drops approximately exponentially towards the interior of the metal: one may even define an attenuation length. For an accurate circuit theory this attenuation will take place *within* the black box description. A consistent circuit theory can now be obtained if the size of the metals, so also the sizes of possible metallic islands, is large compared to (a) the De Broglie-wavelength and (b) the attenuation length.

Finally, let's discuss the energy conservation in the circuit for these specific tunneling events (the uppermost arrow). It is clear from figure Fig. 8.9 that the hot electrons must lose energy. In this case each hot electron dissipates an amount of eV_s upon arriving at the Fermi level at the positive side. The energy provided by the source to bring an electron to the Fermi level at the negative side is also eV_s and, thus, energy is conserved.

8.4.3   *Filling empty states at the negative side*

Secondly, consider an electron *not* at the Fermi level that tunnels, for example, one of the arrows in the middle in figure Fig. 8.9. The electron is

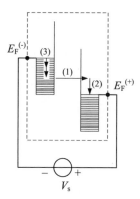

Fig. 8.10   Energy diagram for tunnelling between two metals with fixed Fermi levels. Three different steps can be distinguished: (1) the actual nondissipative tunnelling through the potential barrier, (2) the hot electron losing energy (dissipation) while "falling" to the Fermi level at the positive side, and (3) the filling of the empty energy level at the negative side by electrons with higher energies. The model could be called a tunnel-fall-and-fill model.

not a conduction electron at the negative side. After tunneling this electron is also a hot electron, but the amount of energy that the electron will lose is less than eV_s. In maintaining the voltage across the junction, however, the source has provided again an amount of energy of eV_s. At first sight it seems that the conservation of energy is violated. The hot-electron model has to be extended. By realizing that conduction electrons move on the Fermi level, it is acceptable to suppose that within our black box the position left over by the electron that tunneled, is filled with another electron from a higher energy level. The position of this latter electron is also filled, and so forth until we arrive at a conduction electron at the Fermi level. The energy that is dissipated by electrons that fill up the various states below their own state until the position of the tunneled electron is filled is exactly the amount of energy that is missing. Consequently such a tunnel-fall-and-fill model dissipates the energy delivered by the source, and energy conservation holds again.

Concluding we describe the tunneling event as a series of actions in this way combining the actual tunneling through the insulator with the redistribution of electrons near the tunneling junction, the process is shown in figure Fig. 8.10. That is: *after tunnelling through the barrier electron(s) move towards and from Fermi levels.* With the same reasoning the energy paradox of the tunnel diode in chapter 2 can be understood in circuit theoretical terms.

One might think that the inclusion of the hot-electron model puts severe restrictions on the speed of the devices, due to the fact that measured attenuation times are often in the order of 10^{-12}s. This will not be the case, we will see soon that the condition for tunneling implies that the tunneling electrons are never "very hot", and attenuation can be a rapid process.

8.4.4 *Transition time*

The above described tunnel-fall-and-fill model makes clear that, in general, there will be time involved in the transition of an electron from the negative side of the black box to the other side of the black box. This time is called the transition time. Only in the case that *both* the electrons tunnel from the top of the Fermi level at the negative side to the top of the Fermi level at the positive side *and* both Fermi levels are lying exactly opposite each other *and* we assume a zero tunneling time, then the transition time is zero.

8.4.5 *Capacitive behavior*

Now the capacitive behavior is discussed. First, it is clear that during charging of the junction—like a capacitor—the Fermi levels continuously change. figure Fig. 8.11 shows the Fermi levels in a junction during charging. Because the capacitive behavior is seen when the junction is in Coulomb blockade, electrons arriving at the junction should not tunnel. There is, however, no good reason for why electrons arriving at the negative side of the junction should not tunnel immediately to the positive side. At the negative side many energy states filled with electrons are facing empty energy states at the positive side, as soon as any non-zero amount of charge is collected at the junction. Another mechanism must play a role.

8.4.6 *An extended hot-electron model*

When discussing the resistive behavior of the tunnel junction when excited by a voltage source the Fermi levels on both sides were considered to be static and the reason for tunneling is the possibility to become a hot electron after tunneling; charging effects could not be taken into account. When discussing the capacitive behavior of the tunnel junction when excited by a current source the Fermi levels are dynamically changing all the time—due to charging—but tunneling will not always occur. A model for the occur-

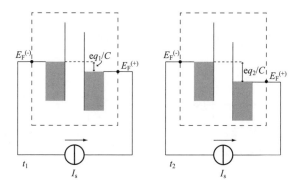

Fig. 8.11   Charging the junction in Coulomb blockade. The Fermi levels shift dynamically and the behavior is like a capacitor. At first sight there seems to be no reason why tunneling of electrons could not occur; many states filled with electrons at the negative side face empty energy states at the positive side. For explaining Coulomb blockade another mechanism must play a role.

rence of the Coulomb blockade should therefore combine the hot-electron model with the possibility of changing Fermi levels.

Consider the consequence of a possible tunnel event in a circuit in which the tunnel junction is excited by a current source I_s. For the moment, we assume that the amount of charge delivered at the junction by the circuit during the tunneling event can be neglected, that is, either the tunneling time is zero either $I_s \times \tau \ll e$, τ being the tunneling time. In figure Fig. 8.12 two situations are sketched: (a) the tunnel junction is charged with a quarter of an electron charge and we consider a possible tunnel event, and (b) the tunnel junction is charged with a little bit more than half an electron charge and a we consider a possible tunnel event (the charge on the metal and in the circuit is considered to be continuous—only bounded currents; only during tunneling the charge is discrete and equal to e). Due to tunneling the Fermi levels will shift, the situations just before and just after tunneling are shown explicitly in the figure. If the junction is charged with $e/4$ Coulomb, figure Fig. 8.12(a), then after tunneling through the barrier the charge on the same side of the barrier will be e/4 - e = (-3/4)e Coulomb and we see that the voltage across the junction is reversed; and so are the Fermi levels. Upon arriving of the complete electron wave the electron "sees" a filled level. If the junction is charged just above e/2 Coulomb then after tunneling the charge on the same side of the junction is just below (-1/2)e Coulomb, and the electron can enter the positive side as a hot electron; consequently, a tunnel event will take place as soon as

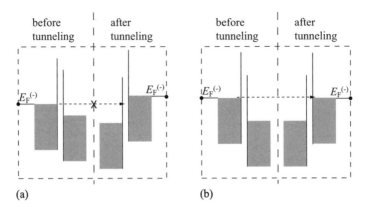

Fig. 8.12   Model for Coulomb blockade and tunneling in a tunnel junction driven by a current source. (a): If the initial charge on the tunnel junction is e/4 the junction shows Coulomb blockade, because the electron (or electron wave) just after leaving the negative side and approaching the positive side experience moving Fermi levels in such a way that the Fermi levels at the negative side appear to be filled. (b): If the initial charge on the tunnel junction is just above e/2 the junction shows a tunnel event, the Fermi levels also change but the electron can still enter the positive side as a hot electron (it enters just above the Fermi level). The difference of the Fermi levels is a bit exaggerated for clarity.

the charge delivered to the junction exceeds e/2 Coulomb.

If we, now, allow the Fermi levels to change dynamically also during the tunneling time then a circuit theory for arbitrary circuits can be obtained. Now, the charge delivered at the junction during the tunnel event has two components: first the charge e of the tunneling electron, and second the charge Δq delivered by the circuit during the interaction between the electron and the barrier: $\Delta q = i \times \tau$, i being the current from the circuit in which the tunneling junction is embedded.

8.5   Tunnel Condition: Mathematical Formulation

The existence of the Coulomb blockade shows that electrons will not always tunnel at a small junction, but that the voltage across the tunnel junction must meet a threshold value, called the **critical voltage**. More precise: *in case of a positive critical voltage tunnelling is possible when the voltage across the tunnel junction is larger than the critical voltage, or, in case of a negative critical voltage tunnelling is possible when the voltage across the tunnel junction is smaller than the critical voltage.* Based on the extended version of the hot-electron model the following formula for the critical voltage is proposed: if the voltage across the tunnel junction just before the tunnel event is $v_{\text{TJ}|\text{before}}$, and the voltage across the tunnel junction just after the tunnel event is $v_{\text{TJ}|\text{after}}$ then for the critical voltage across the tunnel junction holds:

$$v_{\text{TJ}}^{\text{cr}} = \{v_{\text{TJ}|\text{before}} : v_{\text{TJ}|\text{after}} = -v_{\text{TJ}|\text{before}}\}; \qquad (8.19)$$

in words: the critical voltage is the set of all the voltages across the tunnel junction before the tunnel event for which holds that the voltage across the tunnel junction *after* a tunnel event is the opposite of the voltage *before* the event. Using the transit time τ as the time the electron needs to pass from Fermi-level to Fermi-level, we can extend the definition for a non zero tunneling time :

$$v_{\text{TJ}}^{\text{cr}}(t, \tau) = \{v_{\text{TJ}|\text{before}}(t) : v_{\text{TJ}|\text{after}}(t + \tau) = -v_{\text{TJ}|\text{before}}(t)\}. \qquad (8.20)$$

Realize that if the voltage across the junction equals the critical voltage than the transition time reduces to the tunneling time, that is, the time to pass the barrier. At the critical voltage the Fermi level before tunneling at one side is lying exactly opposite the Fermi level after tunneling at the other side.

After formulating the condition for tunneling a general remark is in place. In classical circuit theory any charge is considered to be smeared out, and time is continuous. That is, the number of electrons that contribute to the charge transport in the circuit is such a large number that the individual behavior of a single electron is lost in the contribution of all. When we consider the Coulomb blockade or single-electron electronics in general, however, we find that the tunneling of individual electrons determine the behavior of the circuit. This description, now, causes two different time scales in the circuit. One involves the time within the discrete tunneling took place, modeled by the impulsive current, the other time is associated with the natural response of the circuit.

Problems and Exercises

8.1 Show that the dimension of the delta function $\delta(t)$ is a frequency. (Hint: Use the definition of the delta function.)

8.2 In this chapter tunneling through a potential barrier is considered to be nondissipative and the equivalent circuit for a tunnel event is a charged capacitor parallel with a (delta) current source. What would you suggest as an equivalent circuit if also dissipative tunneling is allowed? Explain why.

8.3 Why is the transition time of a tunneling electron in the case of single-electron tunneling much lower than the average transition time in case of many electrons tunneling in the same time interval?

8.4 What are the reasons to introduce a tunnel-fall-and-fill model?

8.5 When is the transition time equal to the tunneling time?

8.6 Discuss causality in the definition of the tunnel condition, are there really problems? (Hint: what do you know about the voltage across the tunnel junction during tunneling?)

Chapter 9

Impulse Circuit Model for Single-Electron Tunneling—Nonzero Tunneling Times

Based on the black box description and the tunnel condition of the extended hot-electron model, the critical voltage in circuits is determined by a delta pulse current, representing the tunneling electron from Fermi level to Fermi level, a discrete time step τ, the tunneling time, and the voltages across the junction just before *and* after a possible tunnel event, $v_{\text{TJ|before}}$ and $v_{\text{TJ|after}}$ that must have opposite values. With this model we can consider nonzero tunneling times. We will also investigate stochastic behavior of tunnel events that leads to a nonzero transition time.

9.1 SET Junction Excited by an Ideal Current Source— Nonzero Tunneling Time

We understand the tunnel event as an electron sitting at both sides of the barrier at the same time, but the probability of finding the electron going from one side to the other is decreasing on one side of the barrier and increasing on the other side during tunneling. As a consequence of this description charge is moved inside the tunneling junction. Now, also time is involved. The period of interaction between the electron and the barrier is the tunneling time. However, *during* the tunnel event the leads *do not* always supply a current equivalent to the fundamental electron charge divided by the tunneling time. For example, the existence of Coulomb oscillations show explicitly that during the tunnel event not enough current is supplied (if there was, then the oscillations would not be there) [1]. Thus, seen from the outside of the tunneling junction the charge on the junction is changed both by the tunneling electron and the charge delivered by the

---

[1] Apart from this, note that *before* a tunnel event can take place a current is necessary to start a Coulomb oscillation.

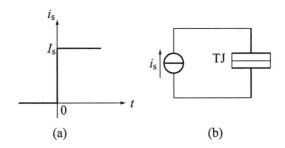

Fig. 9.1   SET junction switched on by an ideal current source: (a) waveform source; (b) circuit

current through the circuit during the time to tunnel.

The total charge on the tunnel junction, now, has three components. First, the amount of charge delivered at the junction before the tunnel event takes place, second, the amount of charge delivered at the junction during the tunnel event, and third, the change of charge by an amount of (-e) Coulomb, e being the elementary charge, due to the actual tunnel event. To describe this last contribution we realize that this change of charge is not associated with a current through the circuit(this argument only holds for bounded currents—not for unbounded currents). In these words discrete (in time and charge) tunneling is combined with continuous charge transfer in the remainder of the circuit.

As a first example the SET junction excited by an ideal current source is considered; the source is switched on at $t = 0$ and has a constant value I_s, see figure Fig. 9.1.

Coulomb oscillations in this circuit can be explained by considering the junction in Coulomb blockade to behave as a capacitor. Due to the current source the junction is charged until the tunnel condition is met. At the tunnel condition the voltage across the junction after tunneling equals minus the voltage across the junction before the tunnel event. In terms of charge, the voltage at the junction before tunneling is q/C and the voltage after tunneling will be changed into $(q - e + \int_\tau idt)/C$; within the parentheses, the first term is the charge on the junction before tunneling, the second term comes from the tunneling electron, and the third term is the contribution of the circuit during the tunneling time and makes it possible to take into account arbitrary circuit elements (including resistors) in the circuitry outside the junction. The only restriction here is that the current must be bounded. Using equation 8.20 we obtain:

$$\frac{(q - e + \int_\tau i dt)}{C} = -\frac{q}{C}. \tag{9.1}$$

Using that $v_{\mathrm{TJ}}^{\mathrm{cr}} = q/C$, the voltage across the junction just before tunneling, we find

$$v_{\mathrm{TJ}}^{\mathrm{cr}}(t, \tau) = \frac{e - \int_\tau i dt}{2C}, \qquad 0 \le \int_\tau i dt \le \mathrm{e}. \tag{9.2}$$

And thus for the constant source I_{s}, $t > 0$:

$$v_{\mathrm{TJ}}^{\mathrm{cr}}(\tau) = \frac{e - I_{\mathrm{s}}\tau}{2C}, \qquad 0 \le I_{\mathrm{s}}\tau \le \mathrm{e}. \tag{9.3}$$

For tunneling time $\tau = 0$, this formula together with the continuously charging and discharging of the junction capacitance shows the phenomenon of Coulomb oscillations with amplitude $e/2C$, as expected. In this case it can be noticed that there is no energy lost during the tunneling event (the Fermi levels are exactly opposite each other before and after the tunnel event), and that the source does not provide energy during tunneling in zero time—but, of course, the source provides energy before and after tunneling that discharges and charges the junction capacitance.

Now, consider the circuit for tunneling times $\tau > 0$. How does the impulse circuit model describe these events? Because of the black-box description we do *not* know what the voltage across the junction is during time τ. What we see from the outside is that a certain critical voltage, $v_{\mathrm{TJ}}^{\mathrm{cr}}$, is reached that marks a start point for electrons to tunnel. And τ seconds later the voltage across the junction is $(-v_{\mathrm{TJ}}^{\mathrm{cr}})$. During the time τ both the electron tunneled and the current source added an amount of charge to the charge initially on the junction, namely $I_{\mathrm{s}}\tau$ Coulomb.

For a given τ, as a consequence of equation 9.2, the amplitude of the Coulomb oscillations will decrease depending on the value of the current source. Figure 9.2 illustrates the situation. The criterion for tunneling is that $v_{\mathrm{TJ|before}}(t) > -v_{\mathrm{TJ|after}}(t + \tau)$. To find the critical voltage we solve:

$$\frac{I_{\mathrm{s}}}{C}t = \frac{e}{C} - \frac{I_{\mathrm{s}}}{C}(t + \tau), \tag{9.4}$$

the voltage immediately after the tunneling of an electron being e/C lower than before the transition. We find

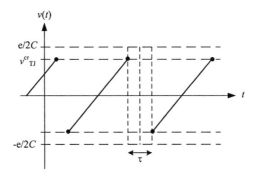

Fig. 9.2   Decreased amplitude if nonzero transition times are considered.

$$t = \frac{e}{2I_s} - \frac{\tau}{2} \qquad (9.5)$$

and for the critical voltage

$$v_{TJ}^{cr} = v_{TJ|before} = v_{TJ}(t) = \frac{I_s}{C}\left(\frac{e}{2I_s} - \frac{\tau}{2}\right) = \frac{e}{2C} - \frac{I_s\tau}{2C}. \qquad (9.6)$$

If we define v_b as the voltage across the junction in absence of a tunnel event and v_a as the voltage across the junction after a possible tunnel event, then graphically the solution can be obtained, see figure Fig. 9.3, by constructing the voltage immediately after a possible tunnel event, formed by the line $v_{TJ}(t) - \frac{e}{C}$, and finding the point for which hold that $v_{TJ}(t + \tau) - \frac{e}{C} = (-v_{TJ}(t))$, that is, $v_{TJ|after}(t + \tau)$ equals $(-v_{TJ|before}(t))$. The reason to look for a graphical solution is that in a later stage the voltages across the tunneling junction for arbitrary circuits are easily found with a circuit simulator, and forms the basis of the calculation of the critical voltage in circuits in the general case.

For large values of the current source the Coulomb oscillations are predicted to disappear (the critical voltage becomes zero). In the latter case, the amount of charge delivered at the junction during the time the electron is tunneling is equal or larger than a single electron charge and consequently the condition for tunneling will be reached before the electron passed the junction; this will result in a continuous current of electrons.

Can the circuit model combined with the tunnel condition still model a tunneling electron, and is energy conserved during the tunneling? At the critical voltage, after time τ the charge on the capacitor is reversed

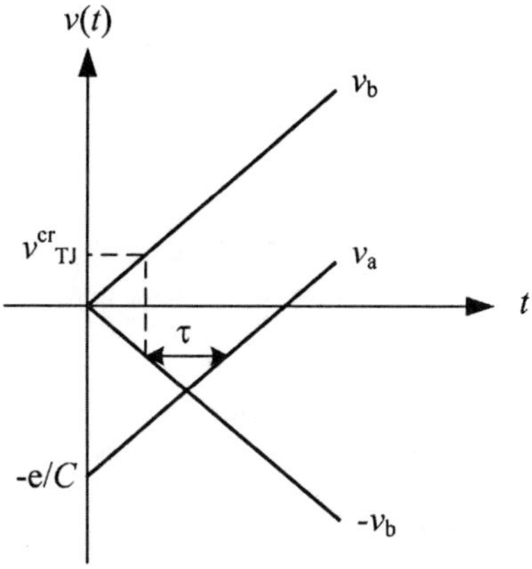

Fig. 9.3   Graphical solution for finding the critical voltage.

and so is the voltage across the junction, the equivalent circuit can only be energy conserving if the sum of the energy necessary for a possible temporal storage of charge and the energy gained by a release of charge and the energy delivered or absorbed by the delta current source is zero during the transition (remember that we do not know the time evolution of the voltage across the junction during the transition). This also guarantees that the constant circuit current source did not deliver net energy during the total transition. To find if it is possible to fulfil these requirements, we consider delta current pulses at various times within the interval t and $t + \tau$. If the delta pulse appears immediately after the critical voltage is reached then during the time τ the source will deliver an amount of charge of $I_s\tau$ but the voltage across the junction (and thus the source) will be negative during the whole interval, this is because the critical voltage, for $\tau \neq 0$ is always smaller than e/2 while the charge contribution of the delta pulse equals e. The source will absorb energy and discharge the junction capacitance. The delta pulse itself will, however, deliver energy because the sum of voltage just before and after is negative. An analogue reasoning for the delta pulse at the end of the interval shows that the source charged the junction capacitance and the delta pulse generates energy. It is not hard to see that at least in case of a constant current source a net contribution of zero is obtained when the delta pulse appears just in the middle of the interval, and in which case the energy of the delta pulse is zero too. The circuit model represents a valid tunneling event. Note that, in the physical description tunneling was smeared out over the whole interval, the circuit

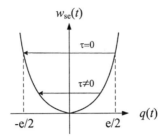

Fig. 9.4 A tunnel junction excited by an ideal current source. Transitions are shown in an energy diagram. The transitions, with and without assuming a transition time (in this case the transition time is only the tunneling time), conserve energy and change the polarity of the voltage across the junction.

description is only a model. However, it is possible to describe a tunnel event and, more important, it is possible to define the expression for the critical voltage.

Because we did not take into account a possible stochastic behavior of the tunneling electron in this section, the tunnel interval τ started immediately after the critical voltage was reached. These tunnel events can also be illustrated as transitions in the capacitor's energy diagram; w_{se} being the energy stored. In figure Fig. 9.4 two transitions are shown. First, when $\tau = 0$ the electron is transported to the other side, by doing so reversing the charge on the junction capacitance. Energy is conserved only when the charge before tunneling is $e/2$ Coulomb. Second, when $\tau \neq 0$ an electron is transported to the other side, but in the mean time the capacitor is charged by an amount $I_s\tau$. Consequently, the net charge difference is smaller than e Coulomb (the electron tunnels to the positive side). Still, energy is conserved before and after the tunnel event, because tunneling is considered to be elastic.

9.2 SET Junction Excited by a Nonideal Current or Voltage Source

The inclusion of resistors in the environment limits the general case to well defined bounded currents through the circuit. Circuits in which the tunnel junction is excited by a nonideal current source or a nonideal voltage source (battery) can be treated as a single entity. The behavior of a circuit with a nonideal voltage source can be obtained from the behavior of the circuit in which the tunnel junction is excited by a nonideal current source, by using

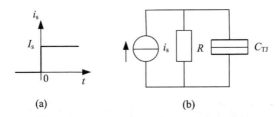

(a)   (b)

Fig. 9.5   SET junction excited by a nonideal current source: (a) source waveform; (b) circuit

source transformation. A series connection of an ideal voltage source V_s and a nonzero finite resistance R is equivalent to a parallel connection of an ideal current source $I_s = V_s/R$, in parallel with R.

In parallel with an ideal current source a resistor (source resistance) is included, a SET junction excited by a real *current* source is obtained. To apply the tunnel condition we need to know the voltage across the tunnel junction before and after a possible tunnel event. The circuit, see figure Fig. 9.5, is described by $I_s = i_R + i_{TJ}$, with $i_R = v_{TJ}(t)/R$ and $i_{TJ}(t) = C(dv_{TJ}(t)/dt)$:

$$C\frac{dv_{TJ}(t)}{dt} + \frac{v_{TJ}(t)}{R} = I_s \qquad (9.7)$$

The solutions depend on the initial charge on and thus on the initial voltage across the junction capacitance v_0 and are well-known:

$$v_{TJ}(t) = v_\infty + (v_0 - v_\infty)\exp(-\frac{t}{RC}), \quad t \geq 0, \qquad (9.8)$$

with $v_\infty = RI_s$ and

$$i_{TJ}(t) = (I_s - v_0/R)\exp(-\frac{t}{RC}), \quad t \geq 0. \qquad (9.9)$$

Again, the junction capacitance will be charged until the critical voltage is reached and tunneling of an electron will occur.

First consider the circuit for a tunneling time $\tau = 0$. Using equation 9.2 and $\tau = 0$ we have:

$$v_{TJ}^{cr} = \frac{e}{2C} \qquad (9.10)$$

and we observe Coulomb oscillations as soon as the voltage across the tunnel junction exceeds the critical voltage, that is, for $|I_sR| > e/(2C)$, that is,

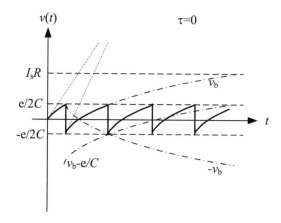

Fig. 9.6   Tunneling in the circuit in which the SET junction is excited by a real current source if the critical voltage is smaller than the steady state value of a non-tunneling circuit.

$|I_s| > e/(2RC)$. For $\tau = 0$ and $e/2C < I_sR$ tunneling is shown in figure Fig. 9.6.

When the voltage across the junction *never* exceeds the critical voltage the junction is in Coulomb blockade; this happens for $|I_sR| < e/(2C)$, that is, $|I_s| < e/(2RC)$. Both figure Fig. 9.7 and figure Fig. 9.8 illustrates why the junction must be in Coulomb blockade. In figure Fig. 9.7 the voltage before tunneling lies on the line v_b, it can be seen that after a (im)possible tunnel event the voltage, that is for example $v_a(t_1)$ or $v_a(t_2)$, is always larger in absolute values than before the event. The condition for tunneling will never be met. In figure Fig. 9.8 is shown that it is not possible to find an intersection point of the curves $(-v_b)$ and $v_b - e/C$ (the points corresponds to impossible tunnel events at $t = t_1$ and $t = t_2$).

The above discussion immediately shows that in case of a real *voltage* source Coulomb blockade can exist too. As was found in the experiment described in section 2.2.3. For the analysis of this circuit, the series combination of an ideal voltage source, a resistor, and the tunnel junction, we transform the circuit into its Norton equivalent by source transformation first. The analysis of the transformed circuit is straightforward.

Now consider the general case allowing tunneling times $\tau \neq 0$. We have:

$$v_{\mathrm{TJ}}^{\mathrm{cr}}(\tau) = \frac{e - \int_\tau i_{\mathrm{TJ}}\mathrm{d}t}{2C}. \tag{9.11}$$

From the point of view of modeling, following the same reasoning as we did in case of the circuit with an ideal source, an additional assumption

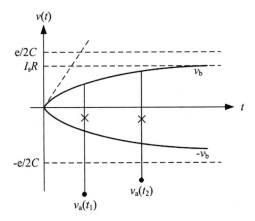

Fig. 9.7 Coulomb blockade in circuit in which the tunnel junction is excited by a non-ideal current source and a zero tunneling time.

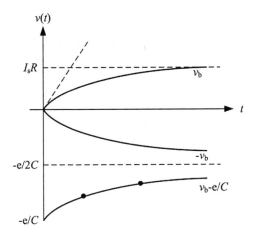

Fig. 9.8 No intersection point in the graphical solution showing Coulomb blockade.

is necessary. Because the current from the circuit during tunneling i_{TJ} depends on the initial voltage across the junction, tunnel events at different times (but within τ) give different initial conditions, and thus different currents and different critical voltages. The energy conserving transition according to a circuit theory is when the delta current pulse takes place at the hypothetical $\tau = 0$ point. Figure Fig. 9.9 shows the predicted behavior.

To summarize the behavior, in general the SET junction excited by a

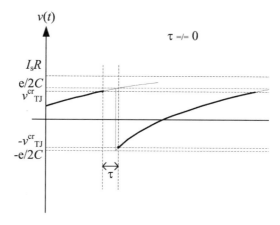

Fig. 9.9   Coulomb oscillations in a circuit in which the SET junction is excited by a real source, when the tunneling time is nonzero.

real source shows:

(1) Coulomb blockade, when the voltage across the junction will never exceed the critical voltage belonging to the circuit.
(2) continuous tunneling, when the critical voltage equals zero, that is, for currents that satisfy the equation $\int_{\tau} i_{TJ} dt = e$, and for larger currents, and
(3) Coulomb oscillations, in all other cases.

9.3   Tunneling of Many Electrons, Stochastic Tunneling, and Resistive Behavior

The description of tunneling of *many* electrons in the same time slot leads to resistive behavior. Stochastic behavior will also lead to dissipation. Both can be modeled by a (tunneling) resistor. When we consider a real energy source, due to the stochastic behavior of tunneling the possibility raises that the actual tunnel event does not take place immediately after the critical voltage is reached, but after a certain *wait time*, sometimes called the dwell time. Because the voltage across the junction will rise above the critical voltage during the wait time: $|v_{TJ|\text{before}}| > |v_{TJ|\text{after}}|$. The difference in Fermi levels before and after tunneling was examined in the case of a junction excited by an ideal voltage source on the level of the individual tunneling electron. There we modeled energy dissipation. We will see now that in this case the calculation of the current produced by the uncorrelated tunneling of many individual electrons in their own time interval (assuming a Poisson distribution) leads to a linear resistance. To understand the origin of dissipation due to stochastic behavior in case of a real source we analyze the situation in the simplified case of zero tunneling time. Note carefully that in this last case we cannot just calculate the energy dissipated in the equivalent circuit of the tunneling junction: $w_{\mathrm{TJ}} = \frac{e}{2}(v_{\mathrm{TJ|before}} + v_{\mathrm{TJ|after}})$, which will be positive, as a measure for dissipation—because for such an individual electron tunneling event the source doesn't provide any energy (we have a bounded current and zero tunneling time) and energy must be conserved.

9.3.1 The probability that a single electron tunnels in a time interval $\Delta\tau$

The calculation of the current produced by the uncorrelated tunneling of many individual electrons in their own time interval (assuming a Poisson distribution) leads to a linear resistance. First consider the classical description of stochastic tunneling or shot noise. We start by determining the probability $P(K, \tau)$ that exactly K electrons tunnel during a time interval of length τ, following [Davenport and Root (1958)]. It seems reasonable to assume: (1) the probability of tunneling during a given interval is statistically independent of the number of electrons that tunneled previously; and (2) that this probability varies as the length of the interval for short intervals, i.e., as $\Delta\tau \to 0$

$$P(1, \Delta\tau) = a\,\Delta\tau, \tag{9.12}$$

where a is as yet an undetermined constant; and (3) that the probability is negligible that more than one electron is tunneling during such a short time interval (electrons than tunnel one-by-one, each in their own interval $\Delta\tau$). Thus approximately for small $\Delta\tau$:

$$P(0, \Delta\tau) + P(1, \Delta\tau) = 1. \tag{9.13}$$

The probability that no electrons are tunneling during an interval of length τ may be determined as follows: consider an interval of length $\tau + \Delta\tau$ to be broken up into two subintervals, one of length τ and one of length $\Delta\tau$. Since the tunneling of an electron during $\Delta\tau$ is independent of the number of electrons that tunneled during τ, it follows that:

$$P(0, \tau + \Delta\tau) = P(0, \tau)P(0, \Delta\tau). \tag{9.14}$$

If we substitute in this equation for $P(0, \Delta\tau)$ the results from equations 9.12 and 9.13

$$
\begin{aligned}
P(0, \Delta\tau)P(0, \tau) &= (1 - P(1, \Delta\tau))P(0, \tau) \\
&= (1 - a\Delta\tau)P(0, \tau) \\
&= P(0, \tau) - a\Delta\tau P(0, \tau)
\end{aligned}
$$

we get for small $\Delta\tau$ the result:

$$\frac{P(0, \tau + \Delta\tau) - P(0, \tau)}{\Delta\tau} = -aP(0, \tau). \tag{9.15}$$

In the limit of $\Delta\tau \to 0$, this difference equation becomes the differential equation

$$\frac{\mathrm{d}P(0, \tau)}{\mathrm{d}\tau} = -aP(0, \tau), \tag{9.16}$$

which has the solution

$$P(0, \tau) = e^{-a\tau}, \tag{9.17}$$

where the boundary condition

$$P(0, 0) = \lim_{\Delta\tau \to 0} P(0, \Delta\tau) = 1 \tag{9.18}$$

follows from equations 9.12 and 9.13. For small values of τ, that is for $\Delta\tau$, we can approach the solution equation 9.17 with:

$$P(0, \Delta\tau) = 1 - a\Delta\tau. \tag{9.19}$$

So now we know the probability $P(0, \Delta\tau)$ and, with Eq. refPE2, thus $P(1, \Delta\tau)$ expressed in the still unknown parameter a.

$$P(1, \Delta\tau) = 1 - P(0, \Delta\tau) = 1 - (1 - a\Delta\tau) = a\Delta\tau \tag{9.20}$$

To obtain an expression for a we consider next that the probability that K electrons are tunneling during an interval of length $\tau + \Delta\tau$. Again we may break up that interval into two adjacent subintervals, one of length τ and the other of length $\Delta\tau$. If $\Delta\tau$ is small enough, there are only two possibilities during the subinterval of length $\Delta\tau$: either one electron is tunneling during that interval, or none are. Therefore, for small $\Delta\tau$

$$P(K, \tau + \Delta\tau) = P(K-1, \tau; 1, \Delta\tau) + P(K, \tau; 0, \Delta\tau) \qquad (9.21)$$

Since again the tunneling of an electron during $\Delta\tau$ is independent of the number tunneling during τ, it follows that:

$$P(K, \tau + \Delta\tau) = P(K-1, \tau)P(1, \Delta\tau) + P(K, \tau)P(0, \Delta\tau) \qquad (9.22)$$

On substituting for $P(1, \Delta\tau)$ and $P(0, \Delta\tau)$,

$$\begin{aligned}
P(K, \tau + \Delta\tau) &= P(K-1, \tau)a\Delta\tau + P(K, \tau)(1 - a\Delta\tau) \\
&= a\Delta\tau P(K-1, \tau) + P(K, \tau) - a\Delta\tau P(K, \tau)
\end{aligned}$$
$$(9.23)$$

we find for small values of $\Delta\tau$ that

$$\frac{P(K, \tau + \Delta\tau) - P(K, \tau)}{\Delta\tau} + aP(K, \tau) = aP(K-1, \tau). \qquad (9.24)$$

Therefore as $\Delta\tau \to 0$ we obtain the differential equation

$$\frac{\mathrm{d}P(K, \tau)}{\mathrm{d}\tau} + aP(K, \tau) = aP(K-1, \tau) \qquad (9.25)$$

as a recursion equation relating $P(K, \tau)$ to $P(K-1, \tau)$. Since $P(K, 0) = 0$, the solution of this first-order linear differential equation is:

$$P(K, \tau) = ae^{-a\tau} \int_0^\tau e^{at} P(K-1, t)\mathrm{d}t. \qquad (9.26)$$

If now we take $K = 1$, we may use our previous result for $P(0, \tau)$ to obtain $P(1, \tau)$. This result can then be used in equation 9.26 to obtain $P(2, \tau)$. Continuing this process of determining $P(K, \tau)$ from $P(K-1, \tau)$ gives

$$P(K, \tau) = \frac{(a\tau)^K e^{-a\tau}}{K!} \qquad (9.27)$$

for $K = 0, 1, 2, \ldots.$
The probability that K electrons are tunneling during a time interval of length τ is therefore given by the Poisson probability distribution.

The average number of electrons tunneling during an interval of length τ is

$$E_\tau(K) = \sum_{K=0}^{\infty} \frac{K(a\tau)^K e^{-a\tau}}{K!} = a\tau \qquad (9.28)$$

since the possible number of electrons tunneling during that interval ranges from zero to infinity. If we define $\bar{n} = E_\tau(K)/\tau$ as the average number of electrons tunneling per second, it follows that

$$a = \bar{n} \qquad (9.29)$$

and that $P(K, \tau)$ may be expressed as:

$$P(K, \tau) = \frac{(\bar{n}\tau)^K e^{-\bar{n}\tau}}{K!}. \qquad (9.30)$$

Since the exponential tends to unity as $\bar{n}\Delta\tau \to 0$, for $K = 1$ and small $\bar{n}\Delta\tau$ this equation reduces approximately to

$$P(1, \Delta\tau) = \bar{n}\Delta\tau \qquad (9.31)$$

which checks with equation 9.12. The probability that a single electron is tunneling during a very short time interval is therefore approximately equal to the product of the average number of electrons tunneling per second and the duration of that interval.

The average number of electrons tunneling per second can be used to find an expression for the tunnel current in case of many tunneling electrons:

$$i = \bar{n}e, \qquad (9.32)$$

with e the elementary electron charge. In nanoelectronics literature $\bar{n}$ is also known as Γ.

9.3.2 Tunnel junction excited by an ideal voltage source: many electrons

Again a SET junction, TJ, excited by an *ideal* voltage source with a DC value v_s is considered. You may also think of a circuit with a real source but with a large current such that the junction will always tunnel continuously, this case can be approximated by the ideal source. We are going to describe this circuit from the point of view that many electrons will tunnel within

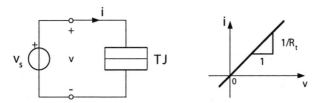

Fig. 9.10 SET junction excited by an ideal DC voltage source: Circuit and predicted behavior at low voltages.

a small time interval. The band diagram of the junction in this circuit shows that many electrons are facing empty energy states, see for example figure Fig. 9.11. Experiments show the resistive behavior of the tunneling junction. Figure 9.10 shows the circuit of a SET junction excited by a DC voltage source and the predicted behavior at low voltages. Today, it is rather easy to perform this tunneling experiment on junctions having large capacitances. The experiments show that for low voltages applied to tunnel junctions the current increases linearly; at higher voltages the current increases exponentially. The ohmic behavior at low voltages, is used to define a tunnel junction resistance R_t.

It was deduced, in section 8.1 that as soon as a *single* electron tunnels the energy delivered by the source must be ev_s, because at the same time the electron tunnels, a current transporting a charge of exactly e (the fundamental charge) has to go through the ideal voltage source in order to keep the voltage across the junction fixed. We found—from the conservation of energy—that there will never be a Coulomb blockade, and electrons will always tunnel towards the positive side of the junction. (If the source has a negative value the electrons will tunnel in the other direction.) All tunneling electrons are hot electrons and dissipate energy until they are part of the thermodynamical electron distribution. A resistive behavior of this continuously tunneling junction is expected. Let's discuss this behavior from the point of view of many tunneling electrons.

At a certain temperature, in case of the tunnel junction excited by an ideal voltage source the actual current i through the tunneling junction is determined by the average number of electrons tunneling per unit time, from the negative to the positive side $\overleftarrow{\Gamma}$, minus the number of electrons that tunnel from the positive to the negative side of the junction $\overrightarrow{\Gamma}$ in the same time span (see figure Fig. 9.11). Based on these definitions the current through the junction is:

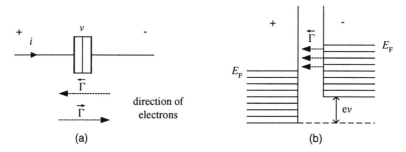

Fig. 9.11   (a) The definition of the tunnel rates; (b) tunnel rate at $T \approx 0$K when the voltage across the junction is fixed (energy diagram showing empty and filled energy states for electrons).

$$i = e \left(\overleftarrow{\Gamma} - \overrightarrow{\Gamma} \right). \tag{9.33}$$

As can be appreciated easily by looking at figure Fig. 9.11 (b), at $T \approx 0$ K, the tunnel rate (thus, the average number of electrons tunneling per unit time) will be proportional to the applied voltage : $\overleftarrow{\Gamma} \propto ev$. In general, the larger the potential across the tunnel junction, the more electrons will be able to tunnel to empty energy states at the other side.

Now, using the known resistive behavior (R_t) and Maxwell-Boltzmann statistics to include the effects of nonzero temperatures, we can express the tunnel rate $\overleftarrow{\Gamma}$ as (see also [Likharev (1999)]):

$$\overleftarrow{\Gamma}(ev) = \frac{ev}{e^2 R_t \left\{ 1 - \exp\left(-\frac{ev}{k_B T} \right) \right\}}, \tag{9.34}$$

where we used that the correct dimensions can be found by, for a moment neglecting the influence of temperature, $i = \Gamma e$ and $v = R_t i$ leading to $\Gamma = (R_t e)^{-1} v$ and thus $\Gamma = (R_t e^2)^{-1} ev$.

Tunneling in the other direction $\overrightarrow{\Gamma}$ is possible if $T > 0$K, and can be found using the Maxwell-Boltzmann distribution:

$$\overrightarrow{\Gamma}(ev) = \exp\left(-\frac{ev}{k_B T} \right) \cdot \overleftarrow{\Gamma}(ev). \tag{9.35}$$

If we approach zero Kelvin ($T \downarrow 0$K) the following expressions are obtained that describe the tunnel junction excited by an ideal voltage source:

$$\overleftarrow{\Gamma}(ev) = \frac{ev}{e^2 R_t} \tag{9.36}$$

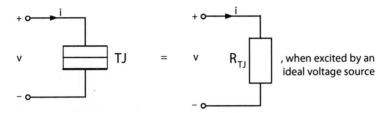

Fig. 9.12   Equivalent circuit of the tunneling junction, when excited by an ideal voltage source; the description is based on the tunneling of many electrons

This is also approximately true in case of $ev \gg k_B T$.

When considering tunneling of many electrons, to describe the equivalent circuit element for the tunneling junction we can use the above discussed behavior. For simplicity, if we consider that $ev_s \gg k_B T$, the theory predicts that the tunnel current will be direct proportional to the voltage across the tunnel junction:

$$i = \frac{v}{R_t}. \tag{9.37}$$

Thus, at this point, the tunneling junction can be described as a linear (passive) resistor circuit element, see figure Fig. 9.12.

9.3.3   *Tunnel junction excited by a current source: stochastic behavior*

In case of a tunnel junction excited by a current source or a nonideal voltage source the effect of a wait time (or dwell time) is shown in figure Fig. 9.13. The voltage across the tunnel junction before tunneling may rise above the critical voltage (the dashed line in the figure shows the tunneling of the first electron that tunnels after the critical voltage had been reached). When looking at the energy-band diagram before and after tunneling, figure Fig. 9.14, we see that this situation can be related to the previous case. After passing the critical voltage the behavior of the circuit changes into a many electron tunneling description and we expect a resistive behavior. A time scale for the delay in oscillations of $\tau = R_t C$ seems to be reasonable.

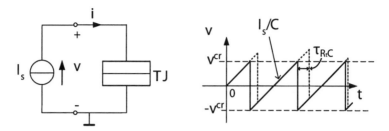

Fig. 9.13　Time delay due to the "probabilistic" nature of tunneling in circuits in which the tunneling junction is excited by a real source that show Coulomb oscillations.

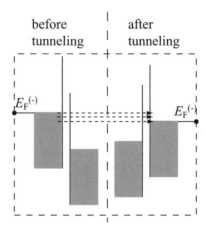

Fig. 9.14　Hot electrons in the description of stochastic behavior in circuits showing Coulomb oscillations

Problems and Exercises

9.1 The inclusion of a tunneling time cannot implemented directly in the impulse circuit model because the delta pulse only exists at $t = t_0$. In a circuit theory the delta pulse is modeled within the tunneling interval. Why is the delta pulse modeled exactly in the middle of the tunneling event in case of a tunnel junction excited by an ideal current source.

9.2 In case of a real current source (describe the source with a Norton equivalent) the amplitude of the Coulomb oscillations is expected to decrease compared to the case of a zero tunneling time. Give an expression of the critical voltage that includes the source resistance.

9.3 Based on energy arguments, explain the existence of a Coulomb blockade when a tunnel junction is excited by a real voltage source.

9.4 Based on energy arguments, explain the absence of a Coulomb blockade when a tunnel junction is excited by an ideal voltage source.

9.5 Explain the appearance and role of the tunnel junction resistance in circuits in which the tunnel junction is excited by a real voltage source in a model that includes a stochastic description of tunneling.

Chapter 10

Generalizing the Theory to Multi-Junction Circuits

In the previous chapters, the single tunnel junction in Coulomb blockade was modeled essentially as a charged (or charging) capacitor. Single-electron tunneling as a Dirac delta pulse. This picture can be extended to more junctions. Therefore, consider a two tunnel junction circuit in Coulomb blockade. The two capacitors create an island in the circuit, that is, a part of the circuit between the two capacitors is electrically isolated from the circuit. If we allow the possibility that one or more electrons had been tunneled through only one of the junctions then the total net charge of the island and its environment will be nonzero and opposite.

The equivalent circuit for two tunnel junctions in series, both in Coulomb blockade is shown in figure Fig. 10.1. The island is represented by two sides of two capacitors and a short-circuit element; through the short circuit element a current can flow distributing the charge on the capacitor plates. The **floating node** between the capacitors can have a net charge. To emphasize this, the floating node is generally called an **island** and the net charge is called the **island charge** $q_i(t)$.

In chapter 1 (section 1.3) we already examined under what conditions the existence of such an island charge can be included in circuit analysis. In general, the requirements for including in circuit theory, either linear or nonlinear, are that Kirchhoff's laws hold. We found that the island charge $q_i(t)$ must be piecewise constant. That is, only if we consider the island charge constant before and constant after tunneling we can model the non-tunneling junction as a (charged) capacitor in a circuit theory. The discontinuous character of the island charge favors the description of the tunnel current with the delta pulse. Because the system describes a nonlinear, however affine-linear, voltage division *and* the tunneling is described in the time domain the full description of multi-junction circuits

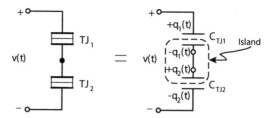

Fig. 10.1   Equivalent circuit for two tunnel junctions in series, both in Coulomb block-ade; the island is represented by two sides of two capacitors and a short-circuit element.

starts in the time domain, instead of in the Laplace domain (which might be easier when dealing only with charged capacitors). The analysis is done for metallic islands with dimensions larger than the electron wavelength. In addition, the absence of stochastic behavior is assumed.

10.1   How Much Energy is Needed to Tunnel onto a Metallic Island?

In nanoelectronics nothing is what it seems to be. We will start to try to answer two simple questions: how much energy is needed to tunnel onto a metallic island? And when has this energy to be supplied? Figure Fig. 10.2(a) shows a small metallic, initially electroneutral, island embedded in an insulating medium. We can view the system as having a single capacitance C_Σ in which the island reflects one electrode of this capacitor and the environment forms the other electrode.

The increase in energy when an electron is added to an island to which already n electrons have been tunneled is called the **single-electron charging energy** $E_{ce}(n)$:

$$E_{ce}(n) = \frac{((n+1)e)^2}{2C_\Sigma} - \frac{(ne)^2}{2C_\Sigma} = \frac{e^2}{2C_\Sigma}(2n+1). \qquad (10.1)$$

For initially uncharged islands the expression for the single-electron charging energy $E_{ce}(n=0)$ is called the *Coulomb energy* E_C:

$$E_C = \frac{e^2}{2C_\Sigma}. \qquad (10.2)$$

Most texts on SET circuits interpret this Coulomb energy as the electrostatic energy barrier felt by the single electron moving onto or from an

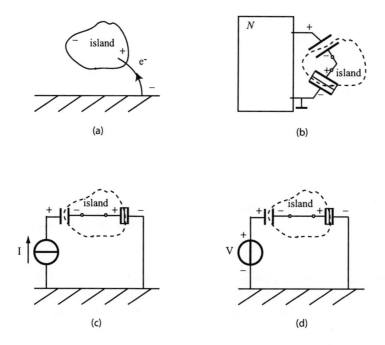

Fig. 10.2 Tunneling of an electron towards an isolated island: (a) non-electric energy source; (b) energy source in electric network N; (c) energy provided by a current source; (d) energy provide by an ideal voltage source.

electrically neutral island. Tunneling is forbidden until this barrier can be surmounted when an energy source is applied that provides enough energy *during* tunneling. In this situation a Coulomb blockade exists if during tunneling not enough energy is supplied.

However, the situation is a bit more complicated, because during tunneling through a tunnel junction energy is locally conserved. We must therefore find out *if and when* there is any energy delivered and, if there is, *where* the energy is stored.

In a circuit theory we model the isolated island by a metallic island surrounded by two metal electrodes. The insulator layer between one of the electrodes and the island is sufficiently small that the island can be charged and uncharged by electrons that tunnel to and from the island. The other electrode forms just a capacitor. The electrodes are connected through an electrical network N; see figure Fig. 10.2(b). The circuit that is formed in this way is called an electron box. We will discuss whether energy is necessary for an electron to tunnel into the electron box. As energy sources we consider first an ideal current source, figure Fig. 10.2(c), and thereafter an ideal voltage source, figure Fig. 10.2(d).

For the following sections, the charge on the capacitor is called q_C, the charge on the tunnel junction capacitance q_{TJ} and we can define the island charge q_i as:

$$q_i = -q_C + q_{TJ}. \tag{10.3}$$

10.2 Electron Box Excited by an Ideal Current Source, Zero Tunneling Time

The current source is switched on at $t = 0$ and has a constant value. Both the capacitor and the tunnel junction are initially uncharged. We start the analysis by finding the Coulomb blockade for the tunnel junction, assuming zero tunneling time.

The current through the circuit will start charging both the junction capacitance and the capacitor. The voltage across the capacitor will increase continuously, the voltage across the tunnel junction will rise until the critical voltage is reached. Upon reaching the critical voltage an electron will tunnel through the tunnel junction. To find the value for the critical voltage we use equation 8.11

$$v_{TJ}(t_{0+}) = -v_{TJ}(t_{0-}). \tag{10.4}$$

Because there will be no distribution of charge within the island in zero tunneling time, the equation is only satified by

$$v_{TJ}(t_{0-}) = \frac{e}{2C_{TJ}}. \tag{10.5}$$

The voltage across the tunnel junction after the tunnel event will be:

$$v_{TJ}(t_{0+}) = -\frac{e}{2C_{TJ}}. \tag{10.6}$$

Consequently the energy dissipated at the tunnel junction, using equation 8.10, is

$$w_{TJ} = \frac{e}{2}[v_{TJ}(t_{0-}) + v_{TJ}(t_{0+})] = 0. \tag{10.7}$$

Also the energy across the capacitor is not changed, because the voltage across the capacitor doesn't change during the tunnel event(remember we consider a zero tunneling time). So it appears that there wasn't any energy increase in energy after the tunnel event!

10.2.1 *Energy considerations: bounded currents*

If the energy in the electron box subcircuit was not increased as a result of tunneling of a single electron, does this mean that there was not any energy necessary to let the electron tunnel, during the tunnel event? To answer this question we look at the energy delivered by the current source (the energy source in the circuit) during the tunnel event. During the tunnel event the energy delivered by the current source is:

$$w_\mathrm{s} = I \int_0^\tau v\mathrm{d}t = 0, \tag{10.8}$$

for $\tau = 0$ and the current in the circuit is always bounded and equal to I.

We see that the current source doesn't supply any energy during the tunnel event too. *During* tunneling there was no energy necessary at all to let the electron tunnel to the island. Of course the source had to provide energy *before* tunneling to charge both initially uncharged junction capacitance and capacitor. Because both capacitances are in series, the energy provided by the current source to charge both capacitances from zero initial charge to the critical value $e/2$ equals the Coulomb energy.

The above described picture is in agreement with the (extended) hot electron model that predicts that tunneling doesn't change the energy of the tunneling junction (energy is conserved during tunneling through the barrier).

10.3 Electron Box Excited by an Ideal Voltage Source

Now, we consider the electron box excited by an ideal voltage source, figure Fig. 10.2(d). In the circuit only consisting of an ideal voltage source and capacitances a delta current pulse may exist when the voltages over the capacitances step as a result of tunneling of a single electron in zero time.

First, we derive some formula. Starting as in section 1.3.1:

$$v(t) = v_\mathrm{C}(t) + v_\mathrm{TJ}(t), \tag{10.9}$$

where $v(t)$ has the value of the constant voltage source V. Now, we include the existence of an island charge (charge on the floating node):

$$q_\mathrm{i}(t) = -q_\mathrm{C}(t) + q_\mathrm{TJ}(t). \tag{10.10}$$

Combining both equations and using the constitutive relation of the capacitor:

$$\begin{cases} v(t) = v_{\mathrm{C}}(t) + v_{\mathrm{TJ}}(t) \\ q_{\mathrm{i}}(t) = -Cv_{\mathrm{C}}(t) + C_{\mathrm{TJ}}v_{\mathrm{TJ}}(t). \end{cases}$$

Solving these two equations with two unknowns, we obtain:

$$v_{\mathrm{C}}(t) = \frac{C_{\mathrm{TJ}}}{C + C_{\mathrm{TJ}}}v(t) - \frac{1}{C + C_{\mathrm{TJ}}}q_{\mathrm{i}}(t), \text{ and} \qquad (10.11)$$

$$v_{\mathrm{TJ}}(t) = \frac{C}{C + C_{\mathrm{TJ}}}v(t) + \frac{1}{C + C_{\mathrm{TJ}}}q_{\mathrm{i}}(t). \qquad (10.12)$$

We also obtain, using the capacitor definition:

$$q_{\mathrm{C}}(t) = \frac{CC_{\mathrm{TJ}}}{C + C_{\mathrm{TJ}}}v(t) - \frac{C}{C + C_{\mathrm{TJ}}}q_{\mathrm{i}}(t), \text{ and} \qquad (10.13)$$

$$q_{\mathrm{TJ}}(t) = \frac{CC_{\mathrm{TJ}}}{C + C_{\mathrm{TJ}}}v(t) + \frac{C_{\mathrm{TJ}}}{C + C_{\mathrm{TJ}}}q_{\mathrm{i}}(t). \qquad (10.14)$$

We continue with expressions for the energy stored on the tunnel junction and the capacitor. In *steady state* ($v(t) = V_{\mathrm{s}}$), we can now immediately obtain expressions for the energy stored at the capacitor and the tunnel junction; leaving out the time dependency:

$$w_C = \frac{q_{\mathrm{C}}^2}{2C} = \frac{C}{2(C + C_{\mathrm{TJ}})^2}(C_{\mathrm{TJ}}^2V_{\mathrm{s}}^2 - 2C_{\mathrm{TJ}}V_{\mathrm{s}}q_{\mathrm{i}} + q_{\mathrm{i}}^2) \qquad (10.15)$$

$$w_{\mathrm{TJ}} = \frac{q_{\mathrm{TJ}}^2}{2C_{\mathrm{TJ}}} = \frac{C_{\mathrm{TJ}}}{2(C + C_{\mathrm{TJ}})^2}(C^2V_{\mathrm{s}}^2 + 2CV_{\mathrm{s}}q_{\mathrm{i}} + q_{\mathrm{i}}^2) \qquad (10.16)$$

Adding these two equations we find for the total stored energy w_{se}:

$$w_{\mathrm{se}} = \frac{CC_{\mathrm{TJ}}^2 + C^2C_{\mathrm{TJ}}}{2(C + C_{\mathrm{TJ}})^2}V_{\mathrm{s}}^2 + \frac{1}{2(C + C_{\mathrm{TJ}})}q_{\mathrm{i}}^2 \qquad (10.17)$$

and for the change in stored energy before and after the tunnel event:

$$\Delta w_{\mathrm{se}} = \frac{1}{2(C + C_{\mathrm{TJ}})}(q_{\mathrm{i}|\mathrm{after}}^2 - q_{\mathrm{i}|\mathrm{before}}^2) \qquad (10.18)$$

If we now define q_{s} as the charge that is transported, by a current $i = q_{\mathrm{s}}\delta(t)$, through the circuit during the tunnel event then:

$$w_{\mathrm{s}} = \int V_{\mathrm{s}}q_{\mathrm{s}}\delta(t)\mathrm{d}t = V_{\mathrm{s}}q_{\mathrm{s}} \qquad (10.19)$$

is the energy provided by the source during tunneling.

Now, we calculate the critical voltage for the electron box in case of unbounded currents, that is, excited by the ideal voltage source. Again we use

$$v_{\mathrm{TJ}}(t_{0+}) = -v_{\mathrm{TJ}}(t_{0-}).$$

Distribution of charge takes place at $t = t_0$ by the current pulse $i = q_s\delta(t_0)$. Considering an electron tunneling to the uncharged box we have a critical voltage for the source V_s^{cr} for which the tunnel condition is met, using equations 10.11 and 10.12:

$$\frac{C}{C + C_{\mathrm{TJ}}}V_s^{\mathrm{cr}} - \frac{e}{C + C_{\mathrm{TJ}}} = -\frac{C}{C + C_{\mathrm{TJ}}}V_s^{\mathrm{cr}}. \tag{10.20}$$

Leading to

$$V_s^{\mathrm{cr}} = \frac{e}{2C} \tag{10.21}$$

and the critical voltage across the tunnel junction becomes

$$V_{\mathrm{TJ}}^{\mathrm{cr}} = \frac{C}{C + C_{\mathrm{TJ}}}V_s^{\mathrm{cr}} = \frac{e}{2(C + C_{\mathrm{TJ}})}. \tag{10.22}$$

To find the current in the circuit during tunneling we can use the difference of the charge at the positive side of the capacitance, because after the tunnel event this amount of charge will be reduced by the amount of charge that flew through the source, using 10.14:

$$q_s = q_C|_{\mathrm{before}} - q_C|_{\mathrm{after}} = \frac{C}{C + C_{\mathrm{TJ}}}e. \tag{10.23}$$

And the energy delivered by the source during tunneling being:

$$w_s = \frac{C}{C + C_{\mathrm{TJ}}}eV_s^{\mathrm{cr}} = \frac{e^2}{2(C + C_{\mathrm{TJ}})} \tag{10.24}$$

and equals the Coulomb energy.

10.3.1 *Energy considerations: unbounded currents*

To understand that the above solution is the *only* solution that satisfy the Kirchhoff laws, energy conservation in the circuit must be checked. We saw that the energy delivered by the source during tunneling is the Coulomb energy

$$w_s = \frac{e^2}{2(C + C_{TJ})}.$$

The energy difference between the stored energy on the capacitor and tunnel junction can be found using equation 10.18:

$$\Delta w_{se} = \frac{e^2}{2(C + C_{TJ})}. \tag{10.25}$$

We see energy is conserved. Instead of calculating the energy stored on the capacitances we could have used the series initial condition model for the charged capacitor and charged junction, and calculate the energy absorbed by the stepping voltage sources representing the stepping charge. To calculate the absorbed energy we again use 8.10

$$w = \frac{q_s}{2}[v(t_{0-}) + v(t_{0+})]$$

and find $w_{TJ} = 0$ and $w_C = \frac{e^2}{2(C+C_{TJ})}$. The calculation of the energies shows that the energy delivered by the source is stored only on the capacitor. At the tunnel junction the energy difference is zero. Of course this is because we started by requiring no energy loss at the tunnel junction. In conclusion we can say that in case of energy provided by an ideal voltage source, the energy necessary to store an electron in the electron box (1) is the Coulomb energy, (2) is delivered during tunneling using an unbounded (delta pulse) current, and (3) is stored only at the capacitor side.

10.4   Electron Box Excited by a Current Source, Nonzero Tunneling Time

Upto now we considered a zero tunneling time. We found that in case of a current source as an energy source no charge was distributed during the tunnel event; in case of an ideal voltage source an amount of $q_s = eC/(C + C_{TJ})$ was distributed during the event. A nonzero tunneling time takes into account the amount of charge transported in the tunneling time τ, and can only be modeled by a circuit in which all currents are bounded currents. This limits the description to junctions to those excited by ideal or real current sources (in case of a real voltage source we can use its Norton equivalent). These circuits will just fill the electron box unless the voltage across the tunnel junction never reaches the critical voltage (due to the presence of resistances).

Due to the nonzero tunneling time the critical voltage across the tunnel junction has values between 0 and $e/2C_{TJ}$, depending on the value of the integral $\int_\tau i d\tau$, as described in chapter 9. When the critical voltage approaches zero the amount of charge distributed will approach e. When the critical voltage is $e/2C_{TJ}$ no charge will be distributed at all. In general calculations have to be done with a circuit simulator to find the various voltages and currents before and after the tunnel event.

10.5 Initial Island Charges and Random Background Charges

Until now we considered only charge on the island that were a result of the currents or voltages in the circuit. It is, however also possible that the island charge is determined by initial charge already on the island or, what are called, **background charges**—charges on the island as a consequence of independent charges in the neighborhood of the island and capacitively coupled to it, for example in the substrate. Especially in the early day of SET technology many SET circuits hampered from these background charges. The often appeared at random and were called random background charges.

It is not difficult to include these charges in the models we described so far. We can still use the models and equations if we adapt the expression for the island charge to include initial and background charge by adding the new charge contributions. For example, in our case of the electron box we define the island charge as:

$$q_i = -q_C + q_{TJ} + q_{initial} + q_{background} \qquad (10.26)$$

In term of the impulse model the additional charge can be modeled as initial charge on the capacitances that form the island. As such island charges will create a nonlinear term in the circuit description in the same way initial charges on capacitors do. Just as in the impulse model the charges can be linearized by considering the initial conditions as sources in parallel or serial initial condition models of the capacitances. Clever circuit design can cope with the appearance of random background charge by adding redundancy in the signal or information processing.

Problems and Exercises

10.1 How much energy is necessary to store an electron in an electron box at $t = t_0$, t_0 being the time moment the electron tunnels? (a) If the electron box is excited by a real voltage source, in the absence of stochastic behavior. (b) If the electron box is excited by an ideal voltage source.

10.2 Repeat the last problem but take stochastic behavior into account.

10.3 What could you say about the value of the critical voltage when non-ideal sources and a nonzero tunneling time are assumed.

Chapter 11

Single-Electron Tunneling Circuit Examples

In this chapter we consider some basic SET circuits in detail. The discussion is limited to those circuits that allow an analytical treatment. Besides this it is assumed that: the tunnel time is zero, the initial island charge is zero, no random background charges are present, and the influence of temperature is neglected. Of course, in this way we are only capable of finding a qualitative description. This qualitative description, however, gives a good idea of the behavior of the circuits. We will consider both bounded and unbounded currents. The reason for considering unbounded currents is, although in real circuits these currents will not be present, that they provide easily an analytical treatment and always provide a value for the critical voltage that is between the two extremes in case of bounded currents including a possible tunneling time. The Coulomb blockade phenomenon calculated in circuits with unbounded currents has a qualitative resemblance with circuits having bounded current if resistors (and thus real energy sources) are included. Besides this, it gives results that can be found mostly in the current literature on SET circuitry. Simulators simulating these idealistic circuits exist.

11.1 Electron-Box

Again we consider the electron box structure, figure Fig. 11.1. First, the structure excited by a current source is described. It predicts the behavior when excited by an constant bounded current. To find the values of the voltage across the tunnel junction that allow for tunnel events, the critical voltages, we consider the tunnel events in which one electron tunnels

217

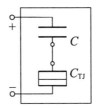

Fig. 11.1   Electron box subcircuit

towards the island on which already n additional electrons are stored:

$$q_i : -ne \to -(n+1)e.$$

11.1.1 *Excited by an ideal constant current source*

In fact this circuit was already described in the previous chapter. Figure Fig. 11.2 shows the behavior. The critical voltage for tunneling through the tunnel junction is $e/2C_{TJ}$. Immediately after tunneling of an electron the current source starts to recharge the tunnel junction; it does this by first discharging the junction capacitance from (-e/2) to zero (realize that the voltage is negative) and second by charging the the junction capacitance from zero to e/2 again. After charging, a next electron can and will tunnel.

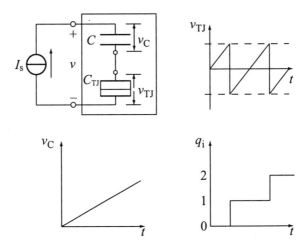

Fig. 11.2   Electron box excited by constant ideal current source; the current through the circuit is bounded.

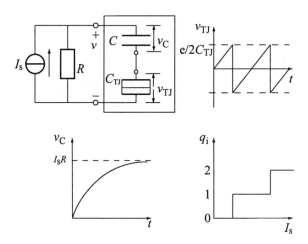

Fig. 11.3   Electron box excited by a real current source; the current through the circuit is bounded.   In the diagram showing the voltage across the tunneling junction is is assumed that $e/2C_{TJ} \ll I_s R$ so that we can assume an almost linear behavior during (de)charging.

In the same time the current source charges continuously the capacitor. The number of electrons in the box will increase continuously in time too.

11.1.2   *Excited by a real constant current source*

When we excite the electron box with a real current source, the circuit is modeled as in figure Fig. 11.3.   The combination of the resistance of the resistor with the capacitance of the capacitor in series with the tunnel junction will lead to a maximal attainable voltage across the electron box: the voltage in steady-state. This maximum voltage determines how many electrons can be stored in the electron box. In this way, in steady-state, the number of electrons in the box is a function of the value of the applied current source.

11.1.3   *Excited by an ideal constant voltage source*

The critical voltage for the electron box in case of unbounded currents, that is, excited by the ideal voltage source, can be found using equations 10.11 and 10.12:

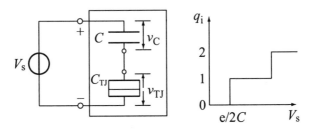

Fig. 11.4   Electron box excited by an ideal voltage source; the current through the circuit is unbounded.

$$v_{\mathrm{TJ}}(t_{0+}) = -v_{\mathrm{TJ}}(t_{0-}).$$

We obtain:

$$\frac{C}{C + C_{\mathrm{TJ}}} V_{\mathrm{s}}^{\mathrm{cr}} - \frac{ne}{C + C_{\mathrm{TJ}}} = -\frac{C}{C + C_{\mathrm{TJ}}} V_{\mathrm{s}}^{\mathrm{cr}} + \frac{(n+1)e}{C + C_{\mathrm{TJ}}}. \qquad (11.1)$$

Leading to

$$V_{\mathrm{s}}^{\mathrm{cr}} = \frac{(2n+1)e}{2C} \qquad (11.2)$$

and the critical voltage across the tunnel junction becomes

$$V_{\mathrm{TJ}}^{\mathrm{cr}} = \frac{C}{C + C_{\mathrm{TJ}}} V_{\mathrm{s}}^{\mathrm{cr}} - \frac{ne}{C + C_{\mathrm{TJ}}} = \frac{e}{2(C + C_{\mathrm{TJ}})}. \qquad (11.3)$$

For each value of the critical voltage an electron will tunnel into the box. As soon as the electron arrives the island charge is raised and the condition for tunneling doesn't hold anymore for this value of the source. Any time a critical voltage is met, exactly one electron tunnels.

We notice a qualitative resemblance of this circuit with the circuit with the real current source (or if we take its Norton equivalent: with a real voltage source). Both circuits show tunneling to the box in discrete steps as a function of the source strength. Only the value of the critical voltage over the tunneling junction is different. This example shows that it is possible the use the analytical power of the solution of the circuit with the unbounded currents to find globally the behavior of the circuits with real sources and bounded currents. Anyway, remember that the inclusion of a tunneling time will also change the critical voltage—to find the critical voltage in these circuits we need the use of a circuit simulator.

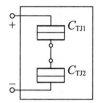

Fig. 11.5   Double tunnel junction subcircuit

11.2   Double Junction Structure

The first modification of the electron box structure is the replacement of the capacitor by another tunnel junction. In this way we obtain a double junction structure, see figure Fig.11.5. To understand its behavior we evaluate two possibilities: either the electron tunnels from the island through junction TJ1, or the electron tunnels towards the island through tunnel junction TJ2.

First consider the tunneling through junction TJ1, while having an uncharged island. When the electron tunnels from the island the island charge becomes e, $q_{i|after} = e$; we perform the analysis using unbounded currents, that is, the subcircuit is excited using an ideal voltage source. The critical voltage is found using:

$$v_{TJ}(t_{0+}) = -v_{TJ}(t_{0-}).$$

We have

$$\frac{C_{TJ2}}{C_{TJ1} + C_{TJ2}} V_s^{cr} - \frac{e}{C_{TJ1} + C_{TJ2}} = -\frac{C_{TJ2}}{C_{TJ1} + C_{TJ2}} V_s^{cr}, \qquad (11.4)$$

giving

$$V_s^{cr} = \frac{e}{2C_{TJ2}}, \qquad (11.5)$$

and the critical voltage across the tunnel junction becomes

$$V_{TJ1}^{cr} = \frac{C_{TJ2}}{C_{TJ1} + C_{TJ2}} V_s^{cr} = \frac{e}{2(C_{TJ1} + C_{TJ2})}. \qquad (11.6)$$

Second, we consider the tunneling through junction TJ2, while the island is charged with charge e. When the electron tunnels to the island the island charge becomes zero again. The critical voltage is found as:

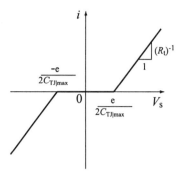

Fig. 11.6   *vi*-chacteristic of the double tunnel junction circuit excited by an ideal voltage source

$$\frac{C_{\mathrm{TJ1}}}{C_{\mathrm{TJ1}} + C_{\mathrm{TJ2}}} V_{\mathrm{s}}^{\mathrm{cr}} = -\frac{C_{\mathrm{TJ1}}}{C_{\mathrm{TJ1}} + C_{\mathrm{TJ2}}} V_{\mathrm{s}}^{\mathrm{cr}} + \frac{e}{C_{\mathrm{TJ1}} + C_{\mathrm{TJ2}}}, \tag{11.7}$$

giving

$$V_{\mathrm{s}}^{\mathrm{cr}} = \frac{e}{2C_{\mathrm{TJ1}}}, \tag{11.8}$$

and the critical voltage across the tunnel junction becomes

$$V_{\mathrm{TJ2}}^{\mathrm{cr}} = \frac{C_{\mathrm{TJ1}}}{C_{\mathrm{TJ1}} + C_{\mathrm{TJ2}}} V_{\mathrm{s}}^{\mathrm{cr}} = \frac{e}{2(C_{\mathrm{TJ1}} + C_{\mathrm{TJ2}})}. \tag{11.9}$$

We see that the critical voltage for both tunnel junctions is the same. In the double junction structure an electron will always tunnel firstly through the junction with the smallest capacitance, because the voltage across this junction will be larger than the voltage across the other one. Immediately after tunneling the voltage across the other junction will be larger than $e/(C_{\mathrm{TJ1}} + C_{\mathrm{TJ2}})$ because of the redistribution of charge, consequently the voltage across this junction is always larger than the critical voltage— $e/2(C_{\mathrm{TJ1}} + C_{\mathrm{TJ2}})$. As soon as an electron tunnels from the island a next one tunnels to the island again. Sometimes this is called correlated tunneling.

If we would have done the analysis starting with the tunneling of an electron to the island, we had found that an electron left the island immediately. From the calculation it is obvious that the structure will conduct as soon as the source voltage meets the critical voltage of the source , $V_{\mathrm{s}}^{\mathrm{cr}}$. This occurs as soon as

$$V_{\mathrm{s}}^{\mathrm{cr}} = \frac{e}{2C_{\mathrm{TJ|max}}}, \tag{11.10}$$

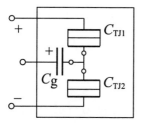 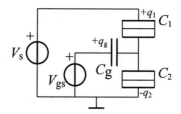

Fig. 11.7   SET transistor subcircuit; SET transistor circuit using ideal voltage sources

where $C_{\text{TJ}|\text{max}}$ is the capacitance of the tunnel junction with the largest capacitance. Figure Fig. 11.6 shows the behavior in the vi-plane; it shows Coulomb blockade below the critical source voltage. Above the critical voltage the double junction structure is described in terms of tunneling of many electrons and is characterized by the tunnel resistance R_t. In case of real sources the structure will also show Coulomb blockade unless the tunneling time involved times the current is sufficiently high.

11.3   SET Transistor

The SET transistor is a device that has two tunnel junctions in series and a gate capacitor on which charge can be controlled by a gate-voltage source. We anticipate that the resulting behavior is the behavior of the double junction structure with a variable critical voltage, because the charge on the gate capacitor will be part of the island charge and as such will influence the charge distribution at both tunnel junctions.

We do the analysis again the case of ideal voltage sources. Writing down the two loop equations (KVL) and the equation expressing the island charge we obtain three equations with three unknowns: V_{TJ1}, V_{TJ2}, and V_g.

$$\begin{cases} V_\text{s} & = v_{\text{TJ1}} + v_{\text{TJ2}} \\ V_{\text{gs}} = v_\text{g} + v_{\text{TJ2}} \\ q_\text{i} & = -q_1 + q_2 - q_\text{g} \end{cases} \tag{11.11}$$

Using the constitutional relation of the capacitor:

$$\begin{cases} V_\text{s} & = v_{\text{TJ1}} + v_{\text{TJ2}} \\ V_{\text{gs}} = v_\text{g} + v_{\text{TJ2}} \\ q_\text{i} & = -C_1 v_{\text{TJ1}} + C_2 v_{\text{TJ2}} - C_\text{g} v_\text{g} \end{cases} \tag{11.12}$$

Using the equations of 11.12, subtracting the second from the first, and adding the third and $(-C_2)$ times the first gives

$$\begin{cases} V_s - V_{gs} = v_{TJ1} - v_g \\ q_i - C_2 V_s = -(C_1 + C_2)v_{TJ1} - C_g V_g \end{cases} \tag{11.13}$$

Solving these equations by multiplying the first equation of 11.13 by $(-C_g)$ and adding it to the second, we obtain

$$V_{TJ1} = \frac{C_g + C_2}{C_1 + C_2 + C_g} V_s - \frac{C_g}{C_1 + C_2 + C_g} V_{gs} - \frac{q_i}{C_1 + C_2 + C_g} \tag{11.14}$$

and

$$V_{TJ2} = \frac{C_1}{C_1 + C_2 + C_g} V_s + \frac{C_g}{C_1 + C_2 + C_g} V_{gs} + \frac{q_i}{C_1 + C_2 + C_g}. \tag{11.15}$$

To find the conditions for tunneling we start calculating $v_{TJ1|after} = -v_{TJ1|before}$ for a given value of the source V_{gs} to find the critical voltage of V_s, V_s^{cr}, for a given V_{gs}. (Having fixed V_{gs} we increase V_s until one of the junctions start to tunnel.) We define

$$C_\Sigma \overset{\text{def}}{=} C_1 + C_2 + C_g \tag{11.16}$$

Assuming that tunneling through junction TJ2 is not possible, consider tunneling through junction TJ1 while the island charge is ne:

$$V_{TJ1|before} = \frac{C_g + C_2}{C_\Sigma} V_s - \frac{C_g}{C_\Sigma} V_{gs} - \frac{ne}{C_\Sigma} \tag{11.17}$$

$$V_{TJ1|after} = \frac{C_g + C_2}{C_\Sigma} V_s - \frac{C_g}{C_\Sigma} V_{gs} - \frac{(n+1)e}{C_\Sigma}, \tag{11.18}$$

for V_s^{cr} holds:

$$\frac{C_g + C_2}{C_\Sigma} V_s^{cr} - \frac{C_g}{C_\Sigma} V_{gs} - \frac{ne}{C_\Sigma} = -\frac{C_g + C_2}{C_\Sigma} V_s^{cr} + \frac{C_g}{C_\Sigma} V_{gs} + \frac{(n+1)e}{C_\Sigma}. \tag{11.19}$$

Solving for V_s^{cr} we get

$$V_s^{cr}(TJ1) = \frac{C_g}{C_g + C_2} V_{gs} + \frac{(2n+1)e}{2(C_g + C_2)}. \tag{11.20}$$

Doing the same calculations for tunneling through junction TJ2 results in

Table 11.1 Expressions for $V_s^{cr}(\text{TJ1})$, tunneling through junction TJ1; and for $V_s^{cr}(\text{TJ2})$, tunneling through junction TJ2; for $n = -1, 0, 1$.

n	$V_s^{cr}(1)$	$V_s^{cr}(2)$
-1	$\dfrac{C_g}{C_g + C_2} V_{gs} - \dfrac{e}{2(C_g + C_2)}$	$-\dfrac{C_g}{C_1} V_{gs} + \dfrac{3e}{2C_1}$
0	$\dfrac{C_g}{C_g + C_2} V_{gs} + \dfrac{e}{2(C_g + C_2)}$	$-\dfrac{C_g}{C_1} V_{gs} + \dfrac{e}{2C_1}$
1	$\dfrac{C_g}{C_g + C_2} V_{gs} + \dfrac{3e}{2(C_g + C_2)}$	$-\dfrac{C_g}{C_1} V_{gs} - \dfrac{e}{2C_1}$

$$V_s^{cr}(\text{TJ2}) = -\frac{C_g}{C_1} V_{gs} + \frac{(1 - 2n)e}{2C_1}. \qquad (11.21)$$

The exprsessions for the critical voltage for the tunnel junctions can be derived, using equation 11.17

$$V_{TJ1}^{cr} = \frac{C_g + C_2}{C_\Sigma} V_s^{cr}(\text{TJ1}) - \frac{C_g}{C_\Sigma} V_{gs} - \frac{ne}{C_\Sigma}$$

$$= \frac{C_g + C_2}{C_\Sigma} \frac{C_g}{C_g + C_2} V_{gs} + \frac{C_g + C_2}{C_\Sigma} \frac{(2n + 1)e}{2(C_g + C_2)} - \frac{C_g}{C_\Sigma} V_{gs} - \frac{ne}{C_\Sigma}$$

$$= \frac{e}{2C_\Sigma} = \frac{e}{2(C_1 + C_2 + C_g)}. \qquad (11.22)$$

We can do the same for V_{TJ2} and will find that the critical voltage across junction TJ2 is the same:

$$V_{TJ2}^{cr} = \frac{e}{2(C_1 + C_2 + C_g)}. \qquad (11.23)$$

In case of an unbounded current in the circuit, we see that the critical voltage across the junctions is a property of the junctions and only depends on the capacitances in the circuit.

We can plot the lines that indicate the critical voltage V_s^{cr}, see figure Fig. 11.8 ($C_2 + C_g > C_1$), the shaded regions show Coulomb blockade. The model describes qualitatively the measurement results on the SET transistor as reported in section 2.2.4. The overall behavior shows a periodicity along the V_g-axis with period e/C_g. It shows that the behavior is the same for different values of additional or missing electrons. The Coulomb blockade around the origin belongs to the blockade if the island charge is zero.

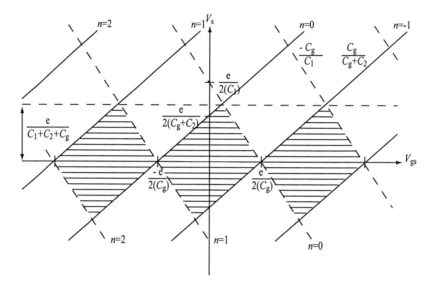

Fig. 11.8   Coulomb blockade in SET transistor circuit using ideal voltage sources. The straight lines indicate tunneling through junction TJ1, the dotted lines indicate tunneling through junction TJ2. The shaded regions are the regions of Coulomb blockade.

Table 11.2   Crossings with the V_s-axis

$$n = 0;\, V_{gs} = 0;\quad V_s^{cr}(TJ1) = \frac{e}{2(C_g + C_2)}$$
$$n = 0;\, V_{gs} = 0;\quad V_s^{cr}(TJ2) = \frac{e}{2C_1}$$
$$n = 1;\, V_{gs} = 0;\quad V_s^{cr}(TJ1) = \frac{3e}{2(C_g + C_2)}$$

The diamond on the left belongs to the blockade if the charge on the island is (-e), $n = -1$; the diamond on the right belongs to the situation that the charge on the island is e, $n = 1$. That the charge on the island can increase and decrease is a consequence of the electron-box structure formed by the gate capacitor on one side and the tunnel junctions on the other side. One of the interpretations of the SET transistor characteristic is that if we take a fixed bias voltage (V_s) in absolute values smaller than $e(2(C_g + C_2))^{-1}$ the transistor can be switched in to and out of Coulomb blockade by switching the gate voltage.

Table 11.3   Crossings with the V_{gs}-axis

$n = 0;$	$V_s^{cr}(TJ1) = 0;$	$V_{gs} = -\dfrac{e}{2C_g}$
$n = -1;$	$V_s^{cr}(TJ1) = 0;$	$V_{gs} = \dfrac{e}{2C_g}$
$n = 1;$	$V_s^{cr}(TJ2) = 0;$	$V_{gs} = -\dfrac{e}{2C_g}$
$n = 0;$	$V_s^{cr}(TJ2) = 0;$	$V_{gs} = \dfrac{e}{2C_g}$

Table 11.4   Slopes and priodicity

slope $V_s^{cr}(TJ1)$	$=$	$\dfrac{C_g}{C_g+C_2}$
slope $V_s^{cr}(TJ2)$	$=$	$-\dfrac{C_g}{C_1}$
periodicity	$=$	$\dfrac{e}{C_g}$

11.4   Three Junction Structure

Circuits consisting of three junction contain two islands. By placing capacitors at both islands the charge at the tunnel junctions can be influenced by two gate voltages. Using a clever voltage clocking scheme this circuit can be used to "pump" electrons through the structure by suppressing the Coulomb blockade of successive tunnel junctions. Figure Fig. 11.9 shows the three junction subcircuit called the electron pump. In the single-electron pump circuit electrons can tunnel through the circuit one by one.

To find a proper clocking scheme describe the characteristics of the subcircuit in a plane with axes V_{gs1} and V_{gs2}, called a state diagram. As a result of the investigations of the SET transistor structure we expect the state diagram to have the following properties: (1) along the V_{gs1} axis the lines indicating a critical voltage show a periodicity of e/C_{g1}, (2) along the V_{gs2} axis the lines indicating a critical voltage show a periodicity of e/C_{g2}, and we expect to have three different slopes each indicating tunneling through one of the three tunnel junctions. We also expect that the slopes will have the same values for different island charges. To find the lines in the state diagram we use the properties above and derive the lines for the critical voltages only for island charges equal to zero. Because in Coulomb blockade the junctions can be represented by linear capacitors and the island charges are zero we can find the actual slopes in the diagram by determining the slopes of the critical voltage expressed in the source values, which can be

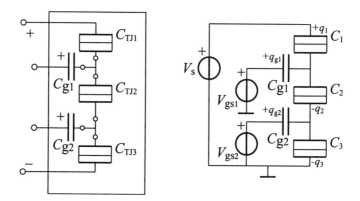

Fig. 11.9   SET pump subcircuit; SET pump circuit using ideal voltage sources.

obtained relatively easily by superposition. The procedure is illustrated by again considering the SET transistor. This procedure is very powerful and can be used to analyze all circuits with unbounded currents in principle.

11.4.1   *Using superposition to calculate the slopes in the state diagram of the SET transistor*

The slopes in the state diagram of the SET transistor can be found by calculating the voltage variation across the tunneling junction as a function of the external sources. For the calculation we use the superposition property of the linear network, as shown in figure Fig. 11.10. Because the island charge is assumed to be zero, the voltage across tunnel junction TJ1 can be written as

$$v_{\text{TJ1}} = av_{\text{s}} + bv_{\text{gs}}. \tag{11.24}$$

For the critical voltage across TJ1 holds that for any value of v_{gs} there is a critical source voltage v_{s} for which holds

$$v_{\text{TJ1}}^{\text{cr}} = av_{\text{s}}^{\text{cr}}(\text{TJ1}) + bv_{\text{gs}}, \tag{11.25}$$

and we find

$$v_{\text{s}}^{\text{cr}}(\text{TJ1}) = -\frac{b}{a}v_{\text{gs}} + \frac{1}{a}v_{\text{TJ1}}^{\text{cr}}. \tag{11.26}$$

The critical voltage $v_{\text{TJ1}}^{\text{cr}}$ is a constant of the junction (in this case having the value $e/(2C_{\sum})$), and we can find a slope in the $v_{\text{gs}}v_{\text{s}}$-diagram by calculating $(-b/a)$. For tunneling through tunnel junction TJ1 we obtain:

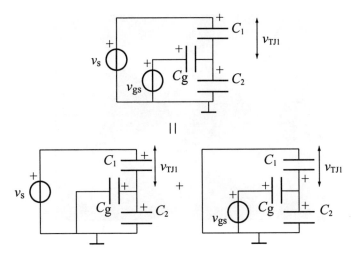

Fig. 11.10 Using superposition to calculate voltages in the SET transistor circuit in Coulomb blockade.

$$v_{\text{TJ1}}^{\text{cr}} = \frac{C_2 + C_{\text{g}}}{C_{\Sigma}} v_{\text{s}}^{\text{cr}}(\text{TJ1}) - \frac{C_{\text{g}}}{C_{\Sigma}} v_{\text{gs}}. \tag{11.27}$$

So, we obtain a slope in the $v_{\text{gs}}v_{\text{s}}$-diagram of

$$\text{slope} \quad v_{\text{s}}^{\text{cr}}(\text{TJ1}) = \frac{-b}{a} = \frac{C_{\text{g}}}{C_2 + C_{\text{g}}}. \tag{11.28}$$

Using the superposition principle a similar calculation can be done for tunneling through tunnel junction TJ2, we obtain:

$$v_{\text{TJ2}}^{\text{cr}} = \frac{C_1}{C_{\Sigma}} v_{\text{s}}^{\text{cr}}(\text{TJ2}) + \frac{C_{\text{g}}}{C_{\Sigma}} v_{\text{gs}}. \tag{11.29}$$

And, we obtain the other slope in the $v_{\text{gs}}v_{\text{s}}$-diagram of

$$\text{slope} \quad v_{\text{s}}^{\text{cr}}(\text{TJ2}) = \frac{-b}{a} = -\frac{C_{\text{g}}}{C_1}. \tag{11.30}$$

11.4.2 *State diagram of the general three junction structure*

Using superposition a state diagrams of the general three junction structure can be obtained. The voltages across the tunnel junctions are a function

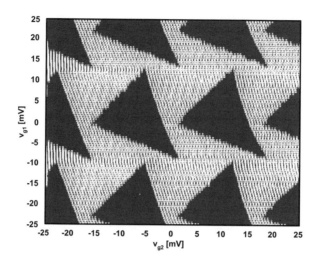

Fig. 11.11   Spice simulation of the electron pump of section 2.2.5.

of the three voltage sources. For example, the voltage across TJ1 can be found as (the subcircuits for the calculation of the parameters b and c can be found in figure Fig. 11.12):

$$v_{TJ1} = av_s + bv_{gs1} + cv_{gs2}, \qquad (11.31)$$

with

$$a = \left(C_{g1} + \frac{C_2(C_3 + C_{g2})}{C_2 + C_3 + C_{g2}}\right) \Big/ \left(C_{g1} + C_1 + \frac{C_2(C_3 + C_{g2})}{C_2 + C_3 + C_{g2}}\right) \quad (11.32)$$

$$b = -C_{g1} \Big/ \left(C_{g1} + C_1 + \frac{C_2(C_3 + C_{g2})}{C_2 + C_3 + C_{g2}}\right) \qquad (11.33)$$

$$c = -C_{g2} \Big/ \left(C_{g2} + C_3 + \frac{C_2(C_1 + C_{g1})}{C_1 + C_2 + C_{g1}}\right). \qquad (11.34)$$

While the formulas for a, b, and c are given explicitly, in general a circuit simulator can be used to obtain the required values. A state diagram is obtained by calculating the critical value of the gate voltage v_{g2} to let an electron tunnel through the tunnel junction under consideration as a function of the gate voltage v_{g1} and the bias voltage v_s. Using the circuit parameters, extracted from the measurements on the electron pump in subsection 2.2.5, a Spice simulation can be performed showing similar behavior, see figure Fig. 11.11.

Example 11.1 As an example, the global behavior of the electron pump (figure Fig. 11.9) is discussed. We consider a simple, but often practical, case that $C_1 \approx C_2 \approx C_3$ and $C_{g1} \approx C_{g2}$ while the junction capacitances are much larger than the gate capacitors.

▽

To get insight in how we can use the gate voltages to manipulate electrons in the structure we take a fixed bias voltage of zero, $v_s = 0$ first. Figures Figs. 11.12, 11.13, and 11.14 show the superposition diagrams before and after simplification.

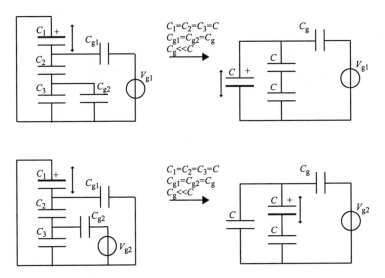

Fig. 11.12   Using superposition to calculate voltages in the electron pump circuit - TJ1.

We see that the simplication results in the following slopes in the state diagram: for tunneling through junction TJ1 the slope is

$$\frac{\Delta v_{g2}}{\Delta v_{g1}} = -2, \tag{11.35}$$

for tunneling through junction TJ2 the slope is

$$\frac{\Delta v_{g2}}{\Delta v_{g1}} = +1, \tag{11.36}$$

and for tunneling through junction TJ3 the slope is

$$\frac{\Delta v_{g2}}{\Delta v_{g1}} = -0.5. \tag{11.37}$$

The state diagram is shown in figure Fig. 11.15. Between parentheses the number

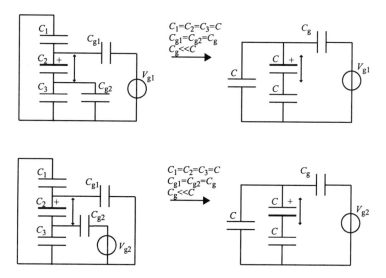

Fig. 11.13   Using superposition to calculate voltages in the electron pump circuit - TJ2.

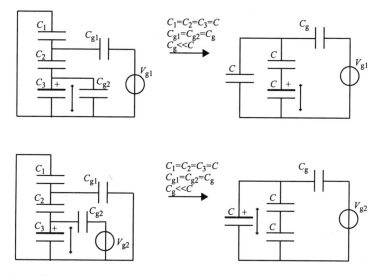

Fig. 11.14   Using superposition to calculate voltages in the electron pump circuit - TJ3.

of additional electrons on the two islands are indicated. The arrows show the pumping principle: applying the correct gate voltages may move an electron from one hexagonal to another, another, and back again, thereby tunneling through all three junctions.

Applying a nonzero bias results in new lines for the critical voltages. With

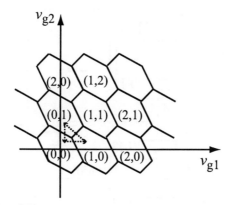

Fig. 11.15 State diagram of a zero-biased electron pump in which $C_1 \approx C_2 \approx C_3$ and $C_{g1} \approx C_{g2}$ while the junction capacitances are much larger than the gate capacitors. The stable regions are hexagonal.

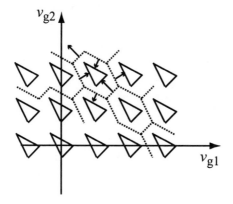

Fig. 11.16 State diagram of an electron pump in which $C_1 \approx C_2 \approx C_3$ and $C_{g1} \approx C_{g2}$ while the junction capacitances are much larger than the gate capacitors. The bias voltage is nonzero.

increasing values for the bias voltage the lines moves as indicated in the hexagonal in the upper right corner. As a result of an increasing bias voltage the shape of the stable regions change into triangles, as is shown in figure Fig. 11.16.

$\triangle$

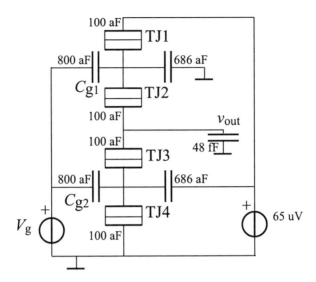

Fig. 11.17   SET-inverter. Due to a large capacitor at the center island, this three island structure is decomposed into two decoupled SET transistors.

11.5   SET Inverter

It is clear that going to structures having even more tunnel junction and islands the analytical tools will not be powerful enough to describe their complex behavior. A general description of a four tunnel junction structure falls outside the context of this introductory text. However, one specific structure will be described briefly. The CMOS-like SET inverter subcircuit consists of four tunnel junctions and, basically, three gate capacitors. The gate capacitor connected to the middle island is in this case much higher than all other capacitors, in this way a structure is created that can be seen as two SET transistors subcircuits in series decoupled by a large capacitor; figure Fig. 11.17 shows the circuit with possible capacitance values. Both SET transistors can be biased such that they never conduct at the same time (as is the case in a CMOS inverter structure). The behavior of the SET inverter similar to the voltage transfer characteristic of the MOS inverter. Figure Fig. 11.18 shows the Spice simulation of the SET inverter of figure Fig. 11.17.

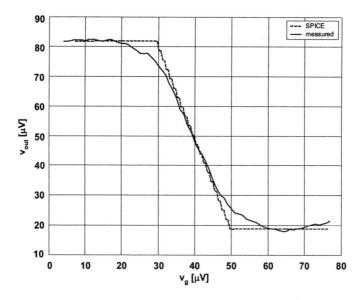

Fig. 11.18 Spice simulation of the SET-inverter circuit.

Problems and Exercises

11.1 Assume zero tunneling time and immediate tunneling (no stochastic behavior). What is the expression for the critical voltage in an electron-box subcircuit when (a) it is excited by current source; (b) it is excited by an ideal voltage source; (c) it is excited by a real voltage source.

11.2 How do the values of the previous problem change if a nonzero tunneling time is assumed?

11.3 How do the values of problem **11.1** change if stochastic tunneling is involved?

11.4 How do the values of problem **11.1** change if room temperature is taken into account?

11.5 Analyze the SET transistor subcircuit when resistors are present representing real source. Assume zero tunneling times.

11.6 What can you say about the values of the gate capacitances in a four junction if (a) the structure is used as a electron pump or (b) as a inverter. Explain the difference.

Chapter 12

Circuit Design Methodologies

Nanoelectronics is still at its infancy. The greater part of texts are discussing physical and the device-physical aspects, and focus on the transport phenomena and basic potential device properties. A minority describe basic electronic circuits, for example [van Roermund and Hoekstra (2000)] and in [Csurgay and Porod (2004, 2007)]. Only few go higher and describe system aspects. The use of nanoelectronics is motivated by several aspects. First of all, the basic devices can be small. Second, it has the potency to operate with very low supply power. Third, the quantum properties that appear at nanoscale in principle represent an increase in signal-processing power. Together, this promises a increase in processing power for future "chips" at very low dissipation. We "just" have to be able to manage it; we will have to exploit the new properties. This vision is opposite, in that sense, to the conventional evolutionary approach that still sticks to the conventional MOS transistor and that tries to counteract the effects that are introduced by the sizing, instead of using them. Both views have in common that they see shrinking down to nanoscale dimensions as inevitable. Having said this, a first introduction to nanoelectronic circuit design methodologies is presented.

12.1 Introduction and Challenges

The main challenge of nanoelectronics research is to exploit the quantum behavior and to solve the problems that we face today: the inherent uncertainties and inaccuracies, the interconnection problem, and the problem how to manage the enormous complexity in the design process.

12.1.1 *Information and signals*

Information is transported by signals, in time and/or space varying quantities (spatio-temporal quantities). In general, in the electronic domain the signals are in time-varying voltages or currents; in the nanoelectronic domain time-varying charge on islands can be added to this. To create voltages, currents, and additional electrons on islands in a nanoelectronic circuit energy sources are necessary. While energy in the circuits is distributed by currents and voltages, and stored in (island) capacitances and inductances. Currents are due to the directed movement of (conduction) electrons under influence of an electric field.

Amplification of signals is necessary. The main reasons for this are the appearance of noise and the loss of signal energy by dissipation, both leading to a decrease in signal-to-noise ratio. Amplification of the primary signal is often the best solution for maintaining a good signal-to-noise ratio. By active amplification of the signal, the signal energy is increased—consequently energy has to be supplied during amplification. This energy is supplied by energy sources in the circuit. Therefore, electronic circuit primitives for amplification have at least three terminals; two for the current through and the voltage over the circuit primitive and one terminal to supply the energy necessary for amplification. Often amplification is nonlinear, in which case only small signals are used and the circuit primitives are described with small-signal models and biasing is used to shift the DC operating point to a well chosen point on the nonlinear transfer characteristic (mostly a point were the characteristic is approximately linear).

12.1.2 *Uncertainties and inaccuracies*

Exploiting quantum effects consequently means: working with devices showing "probabilistic" behavior. This has a great impact on design of nanoelectronics that should not be overlooked. If these quantum effects are to be used, we have to assure that the influence of thermal energy, manifesting itself in thermal vibrations, so in thermal noise, has a negligible effect. In most cases, this means for the moment: working at extremely low temperatures. The challenge for the future is: decreasing the sizes such that quantum energy differences, like the Coulomb energy of an island, become larger than the thermal energy of electrons at room temperature. Low energy levels also involves a high sensitivity to noise which is at the same energy level: every movement of electrons or atoms can be felt; background

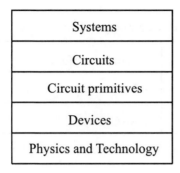

| Systems |
| Circuits |
| Circuit primitives |
| Devices |
| Physics and Technology |

Fig. 12.1 Levels in nanoelectronic system design.

charge is such a notorious noise source. Others are photons and penetrating EM fields. Also the problem of crosstalk between signal lines should not be overlooked with these dimensions.

Sizing goes hand in hand with increased relative inaccuracies, both for absolute and relative accuracy (matching), which introduces substantial inaccuracies in the electronic functions. This is structural, and we will have to cope with that. Further on, design approaches which address this point are discussed. Manufacturability coupled to yield is a serious problem. The challenge here lies in finding new approaches for the production processes. Self-organizing structures, self-repairing, self-assembling, etc., are interesting new ways.

12.1.3   *The interconnection problem*

The basic part of the devices, there where the quantum effects take place, is small. However, if we aim for integration of a large amount of devices, we also have to decrease the interconnection, which in itself is an enormous challenge. One aspect of this is the technological one, that will not be further addressed her. Another one is the architectural one: if we really want to achieve nanoscale circuits, we will have to look for architectures that need a minimum amount of communication via a minimum amount of interconnections, that are mainly local connections. Also the interconnection with the outside world, the I/O is still an unsolved problem.

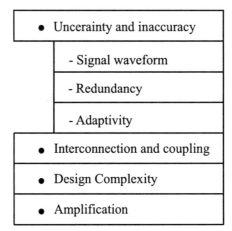

Fig. 12.2   Nanoelectronics-design issues.

12.1.4   *Managing the complexity in the design process*

In nanoelectronics, the design levels, shown in figure Fig. 12.1 are not independent. Often, we cannot split off the system level design from circuit design, and circuit design and device properties are also closely related. The consequence of this is an enormous challenge: how can we (again) come to manageable design processes?

12.2   Nanoelectronic Design Issues

From the foregoing discussion it is clear that there are four classes of problems, see figure Fig. 12.2, the designer of nanoelectronic circuits and systems has to cope with: find a solution to the relative high level of uncertainties and inaccuracies, the interconnection problem, the problem of managing the enormous complexity in the design process itself, and last but not least, find suitable amplification circuit primitives. Discussed are some openings on how to cope with these problems.

12.2.1   *Coping with uncertainty and inaccuracy*

The main part of the threats is caused by the uncertainties and inaccuracies, at different levels. The general way to handle this is to reduce it, or live with it and try to cope with it. Broadly speaking, the relation between output information and input information of any information-processing system is the only thing that matters. Every processing system comprises signals, channels, and noise. The (partial) freedom of the designer is in the definition of the signals, which carry the information, and the definition of the hardware that processes these signals. Suppose, like in this monograph, we want to investigate the feasibility of SET hardware for certain applications (system functions). In that case, the constant factors in the design process are the information to be processed, the technology to be used, and some characteristic properties of the noise sources from the environment. The free factors are: the choice of the signals and the choice of the circuits –and thereby the algorithms– in the SET-hardware. An optimal design means that these choices should be made optimal given the specific properties of the hardware and the noise sources, the constant factors.

Reduction of the uncertainties and inaccuracies looks trivial: first try to make the hardware as accurate enough (as far as that doesn't increase the cost) and try to decrease the noise sources. In fact that is what is called improving technology. *Coping* with uncertainties and or inaccuracies is less trivial, but certainly necessary in nanoelectronics. Three methods are discussed to cope with uncertainties and inaccuracies: the choice of the optimal signal domain(s) where the information is mapped to, the introduction and application of redundancy, and the introduction and application of adaptivity.

- *Choosing the optimal signal waveform and the optimal signal domain.* The information to be processed is translated in two steps into an electrical signal: first it is coded in a dimensionless signal, often a number; second this dimensionless signal is given a physical carrier (voltage, current, number of electrons), resulting in the electrical signal, with value varying in time, that is processed by the hardware. The hardware devices and circuits show strongly different properties for the various domains. Some hardware for instance works good in current domain, some in voltage domain. Or, the properties in the value domain are bad and information can better be put in the time domain. And finally a combination of these is possible. So, the first design choice is the carrier domain. This choice also influences the required biasing of the circuit. Having decided which domain is optimal, the next design issue is how code the information in an optimal way in the signal. Here one might think of all kind of modulation techniques. For example, if the instantaneous value of the output of a SET transistor shows large variations due to background noise, but the amplitude is rather independent of that noise, one might consider putting the information in the amplitude rather than in the instantaneous value. Alternatively, the information could be coded in the number of spikes in a signal that comes out of a SET transistor, or in the periodicity of the spikes, etc. As in nanoelectronics the influence of the imperfections of the hardware is substantial, a good choice deserves considerable attention and plays an important role in the design process and the choice of circuits to be built.

- *Applying redundancy.*

Redundancy literally means an excess of means; sometimes expressed subjectively as superfluous or wasted means, or 'more than strictly seems necessary'. First of all, this can indeed be the case if irrelevant information is processed—which is really wasting means. This should be avoided by proper 'source coding', that is, by proper definition of the signal to be processed. Apart from this, we also can have the situation that the relevant information is transported/processed in a redundant way, for instance several times in parallel or several times successively in time. In electronics we can translate this with: more signal contents and/or more hardware than is strictly required for the implementation of a given function, assuming ideal hardware. At first sight, this looks also a waste of means. However, hardware, including its environment, is never ideal; therefore, redundant hardware is not synonymous with wasted hardware, assuming it is used in a correct way. In addition to this, it can be applied deliberately and with favor to provide the necessary margin to absorb the errors caused by the imperfections of a physical implementation. It was argued earlier that the designer has the freedom to define both signals and hardware, so redundancy can in principle be applied in both. It was also argued that the designer should do this optimally, given the properties of the type of hardware, given the information to be processed, and given the noise from the environment. It is therefore of utmost importance to gain insight in what kind of redundancies can be distinguished and how they can be applied, and, on the other hand, to gain insight in the specific properties of the technology. Here, only the first point is commented upon in further detail, as this is still a general issue that can be applied for all kinds of nanoelectronics technologies.

As said above, the designer has freedom in choosing the signals and the hardware. Correspondingly, we can distinct between two kind of redundancies: signal redundancy and hardware redundancy. Signal redundancy can be explained as follows. After having decided in which electrical domain the processing will take place, the information is mapped on an electrical signal in such a way that there is information redundancy in the signal. This can then be used to absorb the 'errors' caused by the various kinds of noise and system transfer inaccuracies and uncertainties. That means that there is more contents ("information") in the signal than strictly necessary in case of ideal hardware, considering the amount of input information that must be processed. Digital examples are well known, as source coding; the signal is a sequence of discrete-level words, where the number of input information bits is less than the number of bits in the signal. Analog signals too can possess redundancy: the signal, in time and/or in signal value, carries more information than we deliberately put in it, and that extra information will also be processed. In both examples, we have some margin, some surplus, in the signal that grants it extra robustness in the physical implementation, where noise and system-transfer errors are introduced.

Depending on the signal domain(s) chosen, the redundancy can be applied in the time or in the value-domain (voltage, current, etc.), or in both. In principle, the number of possibilities is countless. An example: spreading the information in the time domain followed by processing and subsequently averaging, to obtain a result with more reliability. This can be done continuously or only for discrete events. Another example: discrete information might be coded in the continuous-carrier domain and subsequently processed with a higher resolution than strictly required for the used discretization, enabling signal restoration with a quantizer at the end (a general redundancy used in digital hardware). Instead of coding the information in a single signal, also coding into several signals can be applied (each of them to be represented in the physical system by a separate electrical signal). In case less information is coded in a signal the redundancy in that signal will increase.

Hardware (and algorithmic) redundancy is the second type of redundancy. The physical implementation is now given a surplus of means to make it function more robust in the presence of physical errors caused by noise signals from the environment and imperfections in the hardware. As a result, the functioning of the system will be less vulnerable to inaccuracies in the hardware, so it has a built-in hardware redundancy. This way of introducing redundancy can result in a surplus of hardware and signals, or a surplus in voltage or current range. The surplus of hardware manifests itself in a algorithmic surplus; algorithm and hardware are just different abstraction levels. Note that at higher abstraction levels, the redundancy might be invisible in the system function itself, and only become manifest in its performance. A simple and well-known example of hardware redundancy is used for instance in computer systems in critical situations, like in a space shuttle or nuclear power station, where several computers do the same job; the output given by the majority is considered the right one. Another method is averaging the output of several equal paths. Both are simple forms of hardware parallelism. An alternative method is using the same hardware more times than 'strictly necessary', and combining the results. More complex examples with different paths are found in e.g. fuzzy logic and neural networks, like 'cooperative processing' and a 'winner takes all' processing.

Hardware redundancy might ask for different signals for the various channels, and therefore for coding the input information in a set of different signals. However, this is not necessarily the case; the various channels might also use the same signal. Extra signals on the other hand require extra hardware or hardware operating at higher speed. It is thus clear that there is a relation between signal and hardware redundancy: in both cases we are dealing with redundant processing in an imperfect physical system. Hardware redundancy can be applied at various levels of the design: at device level, at basic circuit level, and at higher system levels. In case of restricted operation ranges, however, like in SET transistor circuits, it seems necessary to apply redundancy already at device or basic circuit level.

- *Applying adaptivity.*

A third way to approach the problem of imperfect hardware is to make your system adapt itself to the desired function, thereby tuning hardware such that the effects of the imperfections are canceled. Seen from a high-level view, this approach makes use of feedback. This approach, complementary to the approach of introducing redundancy, can be a very fruitful one for nanoelectronics. This is the reason why to give attention to neural circuits, as a specific implementation of it. This will be discussed in more detail further on. In principle, this way of tuning can be applied at all levels of the system, so from basic circuit level, up to high system level. However, a general remark has to be made here: tuning can only be effective as long as the tuning parameters stay within their maximum operating range. So, the larger the relative noise and errors, the lower the level where adaptivity must be applied. For nanoelectronics, with relatively low signal levels, and consequently high noise and error levels, this means that the tuning should already be applied rather locally.

12.2.2 *Coping with the interconnection problem*

The interconnection problem is also a severe problem that can be tackled in the technology domain, but also via architectural design. Technology offers several interconnection methods (metal interconnects, substrate connections, etc.) and improvements in technology can lead to e.g. smaller lines and less capacitance to substrate. The architecture gives the two-dimensional topology of the total circuit and thus defines among others if the connections are local or global, if they are crossing each other, and the total number of connections. The less global connections and the less crossings, the faster the circuit can operate. A good architecture, in that sense, is for instance a so-called nearest-neighbor architecture, with ,for example cells in a two-dimensional regular array, or hexagonal honeycomb array, communicating only with their four, respectively, six nearest-neighbors; three-dimensional arrays are even better but are difficult, if not impossible, to realize in an IC. The algorithm performed by the cell array can also be chosen such that the communication between the cells is minimum in the sense of communication speed. In principle almost any architecture can be chosen by the designer, but one has to translate a function into an appropriate algorithm that can easily be mapped on the desired architecture. In fact regularity in an algorithm maps most easily on the regularity of the architecture, and regular functions on their turn map easily on regular algorithms. So the specific function that must be performed also has influence on the final communication. But even in the definition of the function the designer normally has influence: the high-level system behavior can be implemented via various combinations of (sub) functions; e.g. a classification function can be performed by different types of neural networks. The preferable approach to follow here is a bottom up approach to define the architecture, and a top-down approach to map the function via the algorithm on the architecture. In the bottom up phase, defining the architecture, the typical circuit primitive that will be used is of great importance. For SET-circuit primitives for instance, the fan out and the fan in are limited, thereby limiting the number of connections that can be made. The reason for the limited fan in/out is the coupling in the SET circuit primitives between in- and output and the coupling with the capacitive island. So, fan-in and fan-out can be important properties on the basis of which a specific primitive circuit is chosen. Finally, also the I/O to the outside world poses a problem due to the capacitive load and the thermal connection with the environment.

12.2.3   *Coping with the design-complexity problem*

The complexity of some type of nanocircuits, e.g. the SET circuits, is very high due to the non-linear behavior: electron transports only take place if the condition for tunneling is fulfilled and even then the occurrence of such an event follows statistical rules. Between the tunnel events, the state variables are linearly coupled. As the charge on each island corresponds to a state variable, the dimensionality of the circuit is very high. Interpretation of the behavior of a simple SET circuit, like the three-island "inverter" circuit, therefore already becomes very difficult.

A further problem with nanoelectronic circuits is that the levels in the design can not be considered as independent. For instance, two cells interact with each other via their interconnection. In conventional electronics this is prevented by buffering, but in nanoelectronics we try to prevent using buffers, as it would counteract the high density and low power we strive at. Also within cells there is often no full decoupling between in and output. In a SET transistor the drain is still coupled with gate and source.

Structuring is essential for reducing the design complexity. It is a challenge to find a new and efficient classification, a new division into devices, sub-circuits, etc, and to define different models with different abstraction levels that are simple but yet accurate enough to be useful for design. Quantum-classical models, taking into account the statistical behavior are accurate but too much time consuming to be useful for analysis of larger circuits, let alone to be of help in designing circuits. Clearly, this book is a attempt to structure the design-complexity problem in the case of single-electron tunneling circuits.

12.3   SET Circuit Design Issues

In the following, some design issues related to single-electron circuit design are discussed. Again only some openings are discussed on how to cope with the issues. Still, much research and actual designs are necessary to complete the overall picture.

12.3.1   *Signal amplification*

Strictly speaking, signal amplification, amplifies the physical carrier that is used for transporting the information. Therefore, in nanoelectronic SET circuit design we may distinguish voltage amplification, current amplifica-

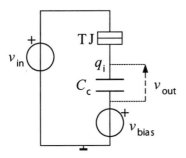

Fig. 12.3 Electron box circuit that can have an amplification larger than one, due to tunneling to the island.

tion, and charge amplification.

When the proposed SET circuits only consist of tunnel junctions, capacitors, and bias sources, while the tunnel condition is not met, the circuit is a passive capacitive circuit and the voltage amplification will be smaller than one always. It is possible that island charges and nonzero bias sources are available in the circuit, but they only contribute to a shift in the transfer characteristic, making it nonlinear. Amplification with a factor larger than one can only achieved if we consider a changing island charge as a consequence of a tunnel event. Due to the tunnel event the transfer characteristic will become nonlinear and opens the possibility of amplification A of small signals, with $A > 1$. We can find the amplification in the voltage domain by considering an electron-box with an addition "small-signal" source v_{in}, see figure Fig. 12.3.

The output v_{out} can be written as

$$v_{out}(t) = \frac{C_{TJ}}{C_{TJ} + C_c} v_{in}(t) - \frac{C_{TJ}}{C_{TJ} + C_c} v_{bias}(t) + \frac{1}{C_{TJ} + C_c} q_i(t). \quad (12.1)$$

Only in those cases in which the island charge is changing by a small input voltage, that is, the tunnel junction is already charged by the bias source with a value just below the critical charge, amplification larger than one can be obtained. In this case it is probably better to speak about charge amplification: the presence of a small fraction of the fundamental electron charge at the junction due to the small-signal source causes the tunneling of an electron, wich in turn leads to a redistribution of charge across the capacitance of a larger fraction (typically, in case of unbounded currents the amount of redistributed charge is $e/2$). By choosing the bias voltage such that the charge on the tunnel junction approaches the critical charge, the amplification approaches theoretically infinity. In practical cases, the amplification is limited due to noise, tunneling times, and resistors in the environment.

12.3.2   *Biasing*

From the previous discussion it is clear that biasing SET circuit primitives in order to precharge tunnel junctions and gate capacitances is of great importance. Small variations in the input may cause single electrons to tunnel and consequently cause "large" voltage variations. It is custom to define an "operating window" around the critical voltage by biasing the circuit primitive. A disadvantage is that we have to know the minimum and maximum values of the input values *a priori*. It is also possible to define multiple windows. The operation on the input is different in different operating windows, as such it is possible to make adaptive functionality; the different operating windows can be selected by different bias values.

12.3.3   *Coupling*

A circuit in SET circuit design is designed by coupling of SET circuit primitives, like those presented in chapter 11. In general, in electronics circuit primitives are represented by two-ports. A two-port has two one-ports, see section 6.1 representing the input and the output, respectively. The input and output quantities are current and voltage. Two requirements hold for the ideal coupling of circuit primitives: circuit primitives should not influence each other's transfer characteristic, and between two circuit primitives only one quantity is chosen as signal for transferring the information. The output signal of one is the input of the next circuit primitive. In practical situations we have to accept a non-ideal coupling with all its consequences.

Preferable, circuit primitives are coupled by ideal coupling blocks. Coupling in nanoelectronics is a major problem, not only due to the fact that ideal coupling blocks are hard to design but also because the signal carrying the information is created by only a few electrons. As a consequence the designed circuit primitives may work well, combining the primitives, however, can cause malfunctioning. Even measuring circuit primitives is often not as easy as it looks at first sight. Some ideas have been proposed for coupling single-electron devices. One of those ideas is to couple the blocks by using "regular" components, like CMOS, shown in figure Fig. 12.4 (a). A disadvantage of this kind of coupling scheme is the current through the gate of

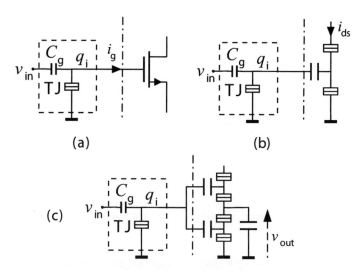

Fig. 12.4   Three examples for coupling of single electron devices. a) Coupling of the SET structure with the help of a CMOS structure. b) Coupling with the help of a C-SET transistor or c) Coupling with a SET inverter.

the transistors (in the figure called i_g). This can be a very low current, but even an extremely low current is still to high for single-electron devices. In those devices it needs to be possible to store a single electron on a node, any current from that node will completely destroy this desired behavior. A second structure proposed to couple the devices is coupling them by a buffer stage build with a single-electron tunneling transistor, shown in figure Fig. 12.4 (b). In spite of its name a current of many electrons (i_{ds}) will flow through such a structure, giving rise to high power dissipation. Also more complex structures, like the SET inverter, can be used—see figure Fig. 12.4 (c).

The solutions are the direct consequence from the standard CMOS design methods. The use of new devices might give rise to new problems, which have to be solved by using different methods. Different functions of the coupling blocks are:

(1) Impedance decoupling: The behavior of the attached building block should not be changed by the coupling scheme used. Because the functionality, like the critical voltage, is mainly influenced by the impedance seen, impedance decoupling is one of the functions which should be included in the coupling scheme.
(2) Bias separation: The functionality of the coupling scheme depends on

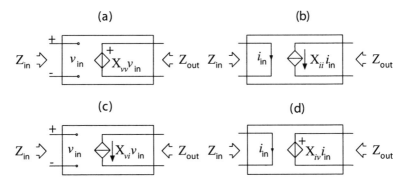

Fig. 12.5   Four implementations of ideal coupling blocks: a) voltage to voltage converter, b) current to current converter, c) voltage to current converter, d) current to voltage converter.

its biasing. To make sure biasing from the previous stage has no influence, bias separation is used.

(3) Conversion: In some cases, when for example the input signals quantity differs from the desired output signal quantity, or when the bias separation or the impedance decoupling is converting the signal from one quantity to another, conversion is needed to obtain the desired quantity.

(4) Directing: To make sure the signal variation at the output is not fed back to the input, some care should be taken in the coupling scheme for directing the signal. An implementation of mainly capacitive components makes the coupling circuit bi-directional.

(5) Amplification: All the separate functions of the coupling scheme described here can cause a weakening of the signal. To make sure the coupling block does not influence the signals amplitude, a certain amplification is needed to get the signal back on its old level.

Ideal coupling blocks.

Assuming two possible input quantities (v and i) and two possible output quantities (v and i) there are four possible ideal coupling blocks. These four ideal coupling blocks are drawn in figure Fig. 12.5.

In general, a coupling block can be defined as:

$$\begin{pmatrix} v_{in} \\ i_{in} \end{pmatrix} = \mathbf{B} \begin{pmatrix} v_{out} \\ i_{out} \end{pmatrix} \tag{12.2}$$

The matrices belonging to the four possible coupling blocks are:

- Voltage to voltage converter (figure Fig. 12.5a)

$$\begin{pmatrix} v_{in} \\ i_{in} \end{pmatrix} = X_{vv} \begin{pmatrix} 1 & 0 \\ 0 & 0 \end{pmatrix} \begin{pmatrix} v_{out} \\ i_{out} \end{pmatrix} \tag{12.3}$$

- Current to current converter (figure Fig. 12.5b)

$$\begin{pmatrix} v_{in} \\ i_{in} \end{pmatrix} = X_{ii} \begin{pmatrix} 0 & 0 \\ 0 & 1 \end{pmatrix} \begin{pmatrix} v_{out} \\ i_{out} \end{pmatrix} \tag{12.4}$$

- Voltage to current converter (figure Fig. 12.5c)

$$\begin{pmatrix} v_{in} \\ i_{in} \end{pmatrix} = X_{vi} \begin{pmatrix} 0 & 1 \\ 0 & 0 \end{pmatrix} \begin{pmatrix} v_{out} \\ i_{out} \end{pmatrix} \tag{12.5}$$

- Current to voltage converter (figure Fig. 12.5d)

$$\begin{pmatrix} v_{in} \\ i_{in} \end{pmatrix} = X_{iv} \begin{pmatrix} 0 & 0 \\ 1 & 0 \end{pmatrix} \begin{pmatrix} v_{out} \\ i_{out} \end{pmatrix} \tag{12.6}$$

The use of these ideal coupling blocks is only needed when a full decoupling is desired. As solution for capacitive decoupling one might use decoupling by using a large load capacitance, like in the inverter circuit primitive; as a solution for directing the use of a current to voltage converter is sometimes suggested. Most of the problems are still unsolved.

12.4 Circuit Simulation

The impulse circuit model complies to the rules of network theory. The device terminals have a voltage v to current i relationship that obeys Kirchhoff's laws under any condition. This model gives a description of the SET junction behavior in local variables and is therefore suitable for standard network theory applications. Because the model parameters are expressed in local variables, it is possible to model the SET junction by circuit simulation tools like Spice.

Such a Spice model is able to operate in a hybrid simulation environment, consisting of SET circuit components in combination with conventional circuit components, which can be very convenient for the verification of measurement results that are in most cases hybrid circuits. A Spice model can be constructed as a direct implementation of the impulse circuit model, except for the shape of the tunnel pulse. The criteria for a tunnel event is determined by the value of the critical voltage, v^{cr} which is a function of the junction capacitance, the current towards the tunnel junction,

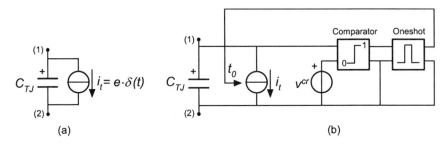

Fig. 12.6   (a) The impulse circuit model. (b) Corresponding Spice model, which is a direct Spice implementation of the impulse circuit model.

and the tunneling time. The amount of charge transported during the tunneling time is the fundamental electron charge, e, in case of zero tunneling time and in case of unbounded currents. In case of bounded currents and a nonzero tunneling time the effective amount of charge transported is $2Cv_{cr}$. For the simulations the following assumptions are made:

- Only one electron e tunnels at the time.
- The critical voltage is equal to half of the voltage drop due to the tunnel event.
- Tunneling is only possible if there is a vacancy for the tunneling electron at the other side of the barrier.

In figure Fig. 12.6, the impulse circuit model and the corresponding (unidirectional) Spice model for a single SET junction are shown. The conditions for unidirectional tunnelling in the impulse circuit model is given by equation 12.7.

$$\begin{cases} v_{\mathrm{TJ}} \geq v^{\mathrm{cr}} : & i_{\mathrm{t}} = e\delta(t - t_0) \wedge v_{\mathrm{TJ}} \geq 0 \\ v_{\mathrm{TJ}} < v^{\mathrm{cr}} : & i_{\mathrm{t}} = 0 \qquad\qquad \wedge v_{\mathrm{TJ}} \in \mathbb{R} \end{cases} \qquad (12.7)$$

Where $\mathbb{R}$ is the set of real numbers. However, in the user environment of Spice, the delta function $\delta(t)$ is not available. To transfer the elementary charge e between the two terminals of the SET junction, a rectangular shaped pulse with a finite width Δt, for which $q = i_{\mathrm{t}} \cdot \Delta t$ holds is chosen as an approximation. In figure Fig. 12.7 this pulse is shown.

The working principle of the Spice model is illustrated in figure Fig. 12.6(b). If the SET junction voltage v_{TJ} exceeds its critical voltage $v_{\mathrm{TJ}}^{\mathrm{cr}}$, the comparator output will be high (which is defined as 1 Volt) and the non-retriggerable one-shot block will generate a pulse with a length Δt and an amplitude which is also chosen 1 Volt. Note that the output signal of

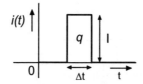

Fig. 12.7 Approximation of the impulse function: a rectangular shaped pulse with a finite width Δt, for which $q = i_t \cdot \Delta t$ holds.

the one-shot functional block is a signal in the time-domain. The starting time is set by t_0 and the width of the pulse is Δt. However, a time signal is not available in the user environment of Spice and therefore time signal t_0 will be represented by v_{t0}.

As long as the one-shot output signal is high, the current source will be active and current will flow so that the tunneling charge q is transported between nodes (1) and (2) of the SET junction, which mimics a tunnel-event. In the single-electron transport regime, the pulse width Δt can be best approximated with 10^{-15} s. The rectangular shaped pulse $i_t = \frac{q}{\Delta t}$ can be implemented with a non-retriggerable one-shot function. This function can be created with programmable Spice building blocks. All other network components of the model can be found in the standard libraries of Spice [is developed at University of California (1975)].

The above described tunnel event is only possible in one direction through the SET junction, when its critical voltage value is met. This means that the user has to know in which direction the SET junction will tunnel, which is very inconvenient. In many cases bidirectional tunnelling is even required for proper operation of the SET circuit and therefore it forms an unavoidable extension to the basic Spice model. One possible implementation to make the basic Spice model bidirectional is to connect the Spice model terminals to a copy of itself in antiparallel. However, the SET junction capacitor C_{TJ} has to be removed from the copy of the SPICE model, because only the tunnel process has to be modeled bidirectional and the SET junction capacitor C_{TJ} is already a bidirectional component. In figure Fig. 12.8 the implementation of the bidirectional Spice model for a single SET junction is shown.

The discussion on the Spice modeling will be concluded by some fine details; for following the main text these may be skipped. The value of one of the critical voltage sources v^{cr} has to be increased slightly with respect to the other source (for example, by less than 1 % of its value) in

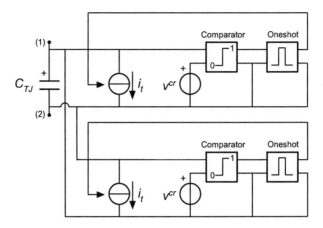

Fig. 12.8   Bidirectional Spice model for a single SET junction. Both basic Spice models are connected in antiparallel to each other.

order to avoid oscillations in the Spice model of the SET junction. These oscillations can occur when v_{TJ} is around the value of v_{TJ}^{cr}. Since the delta function is not available in Spice, a transfer pulse with a certain width Δt was introduced. In some SET circuits, SET junctions can be triggered to tunnel due to a previous tunnel-event, for example in the case in a SET transistor. This means that if this circuit is built using the Spice model discussed, a SET junction can already start to tunnel, even if the previous tunnel-event is not completed. A possible way to avoid these simultaneous tunnel-events is to introduce a general node which keeps track of the status of all SET junctions in the circuit. Every capacitively coupled island in a SET circuit is by definition a 'floating' island. This means that in Spice, the initial charge on such an island is undefined. The initial charge value can be defined with the 'IC=0' statement, or with a two terminal 'IC=0' device. The two terminal 'IC=0' device sets the voltage across itself to 0 V at t = 0s. This two terminal component is shown in figure Fig. 12.9(a). When this device is connected with one terminal to the island, and the other terminal is connected to ground, then the initial charge value is set to zero at $t = 0$, where t represents the simulation start time. This IC=0 device is also inserted into the Spice model of the single SET junction. In this way, all islands created by a voltage source - SET junction loop are automatically taken care of. An example of a voltage source - SET junction loop is given in figure Fig. 12.9 (b), here the the voltage source - SET junction loop consists of v_s, TJ_1 and TJ_2. Such a loop can be distinguished in many

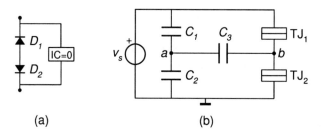

Fig. 12.9 (a) Solution for initial and dynamic charge problems on floating islands. (b) Arbitrary SET circuit showing the floating island problem.

SET (practical) SET circuits. However, if an island is created which is not connected to a SET junction, the user has to place the IC=0 device for this island by hand. The initial charge problem among floating islands is now solved, however, it is still not possible to simulate a SET circuit with a floating node in the Spice environment, because the charge among the islands of a SET circuit are undefined during dynamic simulation runs, like for example a transient simulation. A possible solution to this problem is to place two special diodes in anti-series from every island to ground. These diodes have a certain $i - v$ relation and will form an electrically defined path from island to ground, however, they are not allowed to disturb the functionality of the SET devices. These special diodes can be taken from the DBREAK library. In this library 'ideal' Spice model components can be found. The DBREAK diode has standard a reverse current of 10 fA, but it can be lowered manually. These two special diodes are placed in anti-series to be sure that no diode will conduct at any time.

12.5 Random Background Charges

A major problem in the earlier SET circuitry was the appearance of charge in the substrate that capacitively coupled to the islands. In experiments they were seen as random shifts in the transfer characteristics. They are known as random background charges (RBC). Proper shielding, using a smooth surface (on the atomic level) and a clean substrate (where the number of impurities less than 1 part per million) and using special equipment do help reducing the RBC fluctuations problem, however, they do not avoid them. The RBC problem is not solved completely at the device-level yet, but there are several ways we might try, to deal with this type of disturbance. For example:

- Put the information in the amplitude or frequency component of the signal.
- Use compensation (circuits) to control the charge among the islands.
- Use redundancy on a higher system level in order to reduce the overall error rate.

12.5.1   *Put the information in the amplitude or frequency component of the signal*

Independent of the type of SET circuit, as long as it contains at least one island and one SET junction that shows Coulomb blockade, the transfer characteristic will be periodic. RBC will cause shifts in the characteristic, but the amplitude or frequency will not change. Information coded in amplitude or frequency will not suffer from influence of RBC in those circuits. Both amplitude and frequency can often be controlled by a voltage controlled capacitor coupled to the island. For this solution there is need for an extremely small variable capacitor.

12.5.2   *Use compensation (circuits) to control the charge among the islands*

The compensation techniques are used many times in measurement setups. In many cases an extra capacitor to the island is created in order to tune away the initial island charges and RBC fluctuations that may appear among the particular island by means of external gate voltage sources. This means that every island needs its own compensation capacitor and external (gate) voltage source. In low complexity SET circuits, the island charges could be externally controlled in this way. However, in SET circuits with a large number of islands this (external) option to compensate for unwanted charges, is not feasible.

It is also possible to develop an electronic circuit to compensate for the unwanted island charges, however, if this compensation circuit is made in SET technology it will suffer itself from RBC fluctuations. And if the compensation circuit is made in CMOS technology, the integration density and the low power properties of the SET circuits will be lost. This means compensation is probably not a viable option for SET circuits with a large number of (floating) islands.

12.5.3   *Use redundancy on a higher level (system level)*

Another possible way to overcome the RBC fluctuation problem in SET circuits is to add redundancy to the hardware. This solution does not try to eliminate the problem at the source (as in the case for the previous solutions), but accepts that a particular SET processing element produces a false output. However, more SET processing hardware is needed in order to compensate for the false output for a particular piece of hardware, and obtain the desired overall system functionality.

Redundancy can be applied in many different ways. One possible way to generate redundancy is to use a neural network architecture for the system. In a neural network, different forms of redundancy can be distinguished. The value of the redundancy depends not only on the quality of the neural processing elements, but it is also very dependent on the network topology and if there are the "learning rules" applied. The redundancy is applied to the network in order to obtain robust overall system functionality. If robustness is defined as the allowable amount of faulty neural processing elements with respect to the total number of neural processing elements in the network, in order to remain a valid overall functionality, the robustness can be used as a figure of merit or measure of quality of a neural network. Generally, it will hold that for higher robustness, more hardware must be added. Because, already many buildings blocks for neural nodes have been proposed in SET technologies, this system architecture will be discussed in some detail.

12.6   An outlook to System Design: Fuzzy Logic and Neural Networks

On the basis of the SET circuits we can try to build systems. At this moment, there are problems enough in building the SET circuits, so we are not yet that far that we really can start building systems. However, for a good design in nanoelectronics it is essential to take into account all levels of the design, and to decide at which levels of the design we should cope with the various specific properties of the nanoelectronic circuits. Some general observations on architecture and interconnections have already been made. Briefly, the feasibility of fuzzy logic and neural networks are considered for SET-based electronics. Both have specific properties at information-coding level, signal- processing level, at architecture level. It must be said that there is also a rather strong relation between these properties and the applications we like to be performed by the system; some applications are very well fitted, like pattern recognition and neural networks, other applications might be more difficult to map on them.

12.6.1   *Fuzzy logic*

indexfuzzy logic Fuzzy logic is a good example of how fuzzyness can be used instead of counteracted. It took many, many years before the idea of fuzzy logic, a new paradigm for reasoning, was accepted as a meaningful method for a large number of applications. In its essence, fuzzy logic accepts the fact that information often is only available in a fuzzy way. Based on this recognition, it makes efficiently use of the signal domain: only real information is coded in the signals. Among others, that means that information is not coded uniformly over the whole axis of an input, like temperature; the amount of resolution, so of signal information, is chosen depending on the position on the axis, via classes and membership functions: a large range is just called 'hot', a large just 'cold', and only the essential part is coded via a more accurate membership function of a class 'warm' in a more balanced way. Based on this a complete new type of logic reasoning, called fuzzy logic, has been built, comprising fuzzy algorithms with non-linear functions like maximum or minimum, and several kinds of redundancy via e.g. cooperative processing. Although here the specific coding and the specific logic is based on the 'fuzzy' availability of data, it might also be applied in systems where originally correct data becomes fuzzy due to unpredictable errors, so in systems where the system parameters are fuzzy. It seems a good starting point for building nanoelectronic systems.

12.6.2   *Neural networks*

Neural networks, or better said: artificial neural networks, are networks that try to imitate the behavior of the human brain, which comprises an unlikely amount of very small and basically inaccurate neurons, cooperating in a non-linear way, together performing rather accurately an incredible amount of unbelievably complex functions. Obviously, nature has found a very efficient way of exploiting hardware that differs strongly from the way computers operate. Biology therefore inspires us to follow this unconventional approach, which we shortly will call here the artificial neural network approach, although it comprises far more than what is normally meant with that. Typical properties of human neural networks are the architecture, the

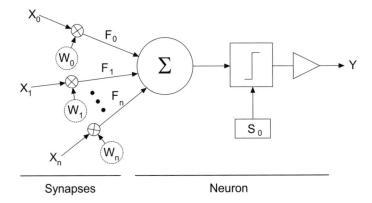

Fig. 12.10   The perceptron, consisting of multiple synapses and a binary neuron.

primitive function performed by a neuron, the type of signal processing, the use of redundancy, and the capability of learning tasks.

Architecture

The architecture of the brain shows a three-dimensional irregular array of neurons with for the greater part nearest-neighbor-like communication via dendrites and axons, coupled via synapses. This is in line with the preference for nearest-neighbor-like architectures to cope with the interconnection problem. The system behavior is partly fixed topologically (the interconnection pattern and the shape and spatial distribution of partially different neurons) and partly electrochemically (e.g. in the coupling weights of the synapses). The topological information is established during growing up, and partly evolutionary determined. The dynamic properties of the components also play an important role in the system behavior. The topological mapping of functions to hardware shows clustering (different classes of functions in different topological parts of the brain), but within a cluster we see functions more or less mapped on common hardware. For nanoelectronics we should also strive for clusters of more or regular structures, with some differentiation between clusters and some, but restricted, communication between these clusters.

- The basic cell

The basic cell is called a neuron. Roughly seen, all neurons look similar; in a more detailed view however, they differentiate at least in size, in shape, and in number of dendrites. So it looks that the principle of similar cells in an array is a good one, but that completely equal cells is not necessary, or even advised against. The way the neurons seem to operate is only

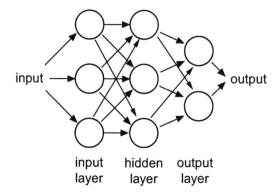

input →

output

input
layer

hidden
layer

output
layer

Fig. 12.11   An example network of connected perceptrons with a hidden layer.

partly known and modeled with different models. One of those models, the perceptron or the McCulloch-Pitts model, is just a very simple model of it, still used by many researchers. It is one of the most well-known neural processing elements for artificial analog neural networks.

The perceptron [Rosenblatt (1962)] is a combination of a neuron and multiple synapses. In figure Fig. 12.10, the complete perceptron is shown. For each perceptron, the following requirements must be met:

- A perceptron must be capable of driving other perceptrons, this means that a buffer-function is required
- The overall signal amplification must be greater than, or equal to 1 in order to avoid signal weakening (which can result in signal loss).

In figure Fig. 12.11 an example of a network of connected perceptrons with a hidden layer is shown. But also the more advanced models are nothing but very simple approaches to the real situation in biology. What first deserves attention is that the processing is analog. This provides a high signal-processing capability, but at the cost of inherent inaccuracies at that level. Obviously nature has found this a good approach, contrary to the digital approach which solves the inaccuracy of the electronics at the primitive cell level (gates). Second: the processing is non-linear, providing also potentially high processing power and means to cope with the inaccuracies. Again, this is in contrast to the conventional analog processing, which operates in a small linearized region. For nanoelectronics it looks favorable to operate in large-signal domain, including non-linearities.

   -Information coding

How information is coded in the brain is not exactly known, but some aspects seem clear. Information is transported via the dendrites and axons more or less as a train of spikes, which shows that information is anyhow coded in the time domain, in number and time position of spikes. But, at least locally, certainly also information is presented by the amplitude of the signals. Also processing takes at least partly place in that amplitude domain. Within neurons the information is presented in the signal-value domain: in the analog values of currents and voltages. The information is further stored in the topology of the system and in the analog weights. Concluding, we can say that biological neural networks use a hybrid form of information coding and that the form chosen at least depends on the properties of the hardware locally. That idea should be applied also in nanoelectronics. It serves in exploiting the hardware optimally.

-Redundancy

There seems a large amount of redundancy be used in biological neural networks, both signal redundancy and hardware redundancy. With respect to signal redundancy, we know that the information from sensors like the eye is splitted and coded in separate signals; all these signals are transported and partly processed in an analog way, so utilizing the full analog information in the signal. At the output of the neuron however, we see the signals being translated to pulses: the neurons 'fire', thereby possibly throwing away signal content in favor of robustness to channel errors. Not necessarily: it might be that all information is completely translated to the time domain. However, this is not very likely if we look to the transport behavior of axons and dendrites, where the information in the signal continuously changes from analog signal-value domain to time domain.

Hardware redundancy is apparent at several levels. Globally we recognize it in the massively parallel operation of neurons, combining their results in non-linear ways, modeled in artificial neural networks with algorithms like 'winner takes all'. But also locally we recognize redundancy, e.g. in the transport mechanism of neurotransmitters within the synapses: many transmitter 'channels' operate there in parallel. On a global algorithmic view the huge amount of neurons together perform a kind of reasoning that also shows large redundancy, as we are also aware from our own way of thinking and remembering. For instance, we know that we use an associative form of memorizing, with a 'surplus' of associations; if we fail in remembering something in one way, our memory will be triggered in another way: there are several ways that lead to Rome. Such an approach in nanoelectronics seems strongly recommended.

-Adaptivity

We have seen that signal coding and redundancy are used extensively. The third method proposed earlier to cope with uncertainties and inaccuracies, viz. adaptivity, is also employed in biological neural networks. It is better known there as learning. The tuned parameters in that case are the weights in the synapses, and the accurate reference that makes it all behave accurate, via feedback, is the 'training vector set', the examples for which we know what the output should be. That is the way we learn e.g. our children to recognize objects. That what is called the supervised way of learning. There is also an unsupervised way: now the reference is a rule, describing a property we require from the output, given a set of inputs. E.g.: the rule that the input vectors should be clustered in a limited number of classes, such that these classes are separated in an optimal way from each other according to a certain definition (rule) of 'distance'.

A nice observation from biological neural networks is further, that various time scales for learning can be exploited, and that that is implemented in different ways. We recognize a short-term learning, with feedback/storage directly via signals (electrically and chemically), medium-term learning established via adaptation of shapes and sizes of synapses and neurons, and long-term learning, reflecting for instance in the topology properties, like the presence or lack of interconnections. Translated to nanoelectronics, this would imply memory functions with different requirements on memory retention times, and different kinds of feedback. Besides the well known electrical feedback of signals on the system function via e.g. multipliers, it must not be overlooked that self structuring based on signal information could be a possible future candidate.

12.6.3   *SET Perceptron examples*

Although not all circuit design ins and outs are known yet, various research groups suggested neural network nodes based on SET circuit primitives and SET devices. As an illustration some of those will be presented briefly. The first circuit ideas date from the end of 1995 [Goossens *et al.* (1995)]. They suggested a neural node consisting of a multiplication function, adding function, and activation function. Central in the design is the SET transistor in series with an current source. A multiplication like function is obtained using this C-SET transistor coupled to two gates, representing the input and the weight. The adder is obtained by capacitive coupling, and the activation function by cascading two SET transistors, see figure Fig. 12.12.

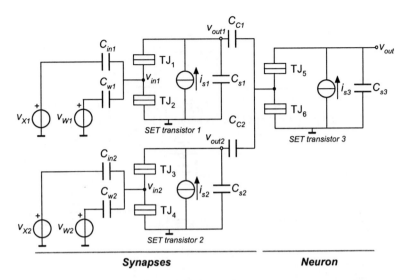

Fig. 12.12 A complete two input perceptron implementation, using three current biased SET transistors.

Another approach is taken in [Kirihara and Taniguchi (1997)]. The perceptron implementation in SET technology shows a complete two input perceptron based on a SET turnstile (a modified electron pump structure), SET inverter 1 and the SET multiplier form the synapse; see figure Fig. 12.13. The dotted box indicates the second synapse. The neuron function is performed by the adder, consisting of coupling capacitors C_{g7} and C_{g8}, and SET inverter 2, which performs the activation function. The weight is represented as a voltage at the drain of the 3 island structure. They also describe a weight storage and in/decrementing device based on an electron pump.

When both complete two input perceptrons are compared the following remarks can be made: first, the latter one is much more complex, has more islands and is therefore more sensitive to disturbances like RBC fluctuations. Second, the first perceptron operates in the relatively high current regime, which means that this neural hardware consumes orders more power than the first, which operates completely in the single-electron transport regime. Figure Fig. 12.14

A last example is a complete perceptron based on SET inverter structures for both the synapses and the neuron operating in the single-electron transport regime by [van de Haar and Hoekstra (2003)]. In the SET inverter

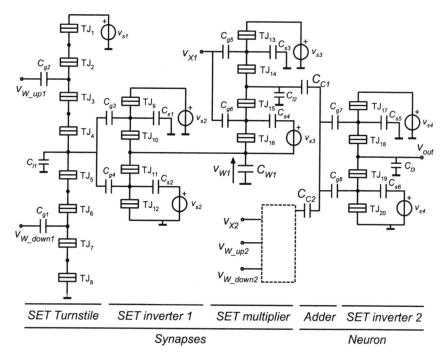

SET Turnstile   SET inverter 1   SET multiplier   Adder   SET inverter 2

Synapses   Neuron

Fig. 12.13   A complete two input perceptron implemented using SET devices. The second synapse is represented by the dotted box.

based synapse structure the weight value v_{W0} is set by the supervisor of the neural system (from a learning system). The digital input signal v_{X0} can be set by the supervisor or by the output of another perceptron. The analog weight value W_i is normalized between 0 and 1. The digital input signal X_i is also normalized and can only be 0 or 1. In order to translate these values to the voltage domain of the SET inverter, the transfer-function with the parameter set as chosen in the figure is taken into account. The load capacitor C_{l1} is set to 480 aF and can hold maximally 20 electrons with this parameter set if input signal X_i is 6 mV. This means that the analog weight value W_i will be discretized to 20 levels. Note that this is a trade-off between the number of allowable voltage levels and the speed (and power dissipation) of the neural hardware. This particular parameter set is chosen in order to obtain good fan-in and fan-out properties.

     In the two input neuron (also based on a SET inverter structure) the summation function is performed by capacitors C_{C0} and C_{C1} and the activation function is performed by the SET inverter structure. This activation

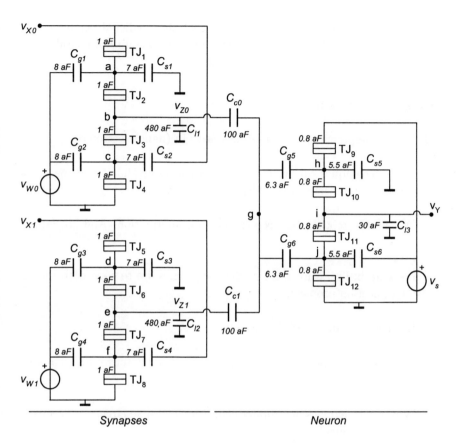

Fig. 12.14 A two input perceptron, based on SET inverter structures.

function is a hard-limiter function. However, the output can yield two extra values, since it is possible to have an electron at islands h and j as well. Thus the output voltage v_Y can be - 0.1 mV, 0.3 mV, 5.9 mV or 6.3 mV around the transition. The parameter set for the neuron is scaled with a factor 0.8 with respect to the parameter set of the synapse, in order to obtain output signals in the same voltage range.

When compared with the previous designs this last one has the lower power consumption of the second and the lesser complexity of the first. All the three designs were simulated either by Spice (the last one) either by the SIMON simulator (a simulator based on the orthodox theory)[Wasshuber *et al.* (1997)]. In all cases the main condition is that the tunneling time is zero; what is not the case in any real application. Besides this resistors could not be included. This all, tells us that the predicted behavior only qualitatively holds. We might say that only the feasibility of neural nodes is shown.

Problems and Exercises

12.1 Why is amplification of signal necessary in (nano)electronics?

12.2 What techniques can we use to cope with uncertainties and inaccuracies in (nano)electronics? Why is this necessary?

12.3 Discuss the relation between signals and information. And how about energy?

12.4 Look up in literature what modern strategies are to cope with the interconnection problem.

12.5 Discuss coupling of SET subcircuits.

12.6 Discuss what is necessary to design a circuit simulator for SET circuits.

12.7 Why are researches developing neural-network-like circuitry for single-electron circuits?

Chapter 13

More Potential Applications and Challenges

The previous chapter showed circuit examples for applications in which redundancy played an important role. This last chapter we will look at some SET circuits and applications for which we assume that perfect manufacturing is possible and initial and/or random background charges do not appear; that is we have an almost perfect technology for creating the devices. Again the simulations are all based on the unbounded tunneling current concept.

13.1 Logic Circuits

Based on the SET inverter structure, described in section 11.5, a logic family can be defined using the property that the structure consists of two serial-connected SET transistors[Klunder and Hoekstra (2004)]. Depending on the input signal only one of the two SET transistors conducts current; the other being in Coulomb blockade. In a way, the structure looks like a CMOS transistor pair. Because of this resemblance, logic gates can be designed based on the known logic gate topologies in CMOS digital circuit design. An example of such a logic structure, an EXOR function, is shown in figure Fig. 13.1

However, basic logic gates can be realized by far simpler SET circuit primitives, using explicitly the tunneling properties of the junction and the existence of island charges. The simplest circuit primitive having these properties is the electron box.

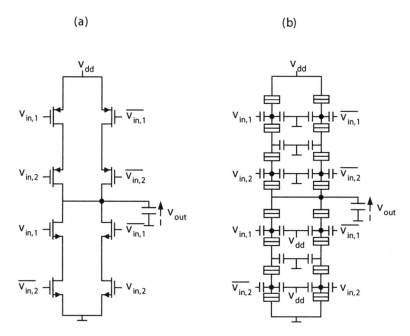

Fig. 13.1 a) The implementation of an EXOR function in standard CMOS. b) The implementation of an EXOR function in SET technology; also the inverse of both input signals is needed.

13.1.1 *Electron-box logic*

When excited by a current source or by a sufficiently large voltage source the voltage across the tunneling junction of the electron box is periodic (see section 11.1). A simulation of the voltage across the tunneling junction due to an applied input voltage is shown in figure Fig. 13.2a. Figure Fig. 13.2b shows the well known staircase of the same electron-box.

The electron-box was treated in chapter 11 where we found that in case of excitation by an ideal voltage source, the period was (equation 11.2):

$$\Delta v_{\text{in}} = \frac{e}{C_{\text{C}}}. \tag{13.1}$$

In figure Fig. 13.2a the periodicity due to tunneling of electrons can be seen. We can use this periodicity to make complex circuitry. Or we can use the input-output relation of figure Fig. 13.2b to build multi-level logic. In the following the design of the NAND, NOR, inverter and threshold function will be discussed. It is enough to consider only the region near the first

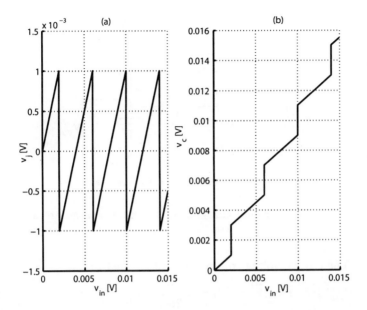

Fig. 13.2 Simulation results of an electron-box. a) The voltage across the tunnel junction v_j as function of the input bias voltage v_{in}. b) The voltage over the normal capacitor, v_C, as function of the input voltage v_{in}. Both simulations are carried out with $C_C = 40aF$, $C_j = 40aF$.

tunnel event. If we use a small signal source v_{in} we will only shift a little away from the event.

Consider the subcircuit shown in figure Fig. 13.3, a more general case of the electron-box, as a programmable block. Let a be the number of inputs. The normal capacitor C_C is divided into $a + 1$ capacitors $(C_{in,1},...,C_{in,a}, C_c)$, which allows us to create various digital building blocks. Let C_j be the capacitance of tunnel junction TJ, and let C_Σ be the sum of the capacitance values attached to node m (the island), $C_\Sigma = C_c + C_j + \sum_{k=1}^{a} C_{in,k}$. We write for the output voltage, the voltage across the tunnel junction, of the electron-box *before the tunnel event*:

$$v_{out}(t) = \frac{v_C(t)C_C + \sum_{k=1}^{a} v_{in,k}(t)C_{in,k}}{C_\Sigma}$$

and *after the tunnel event*:

$$v_{out}(t) = \frac{v_C(t)C_C + \sum_{k=1}^{a} v_{in,k}(t)C_{in,k} - e}{C_\Sigma}. \tag{13.2}$$

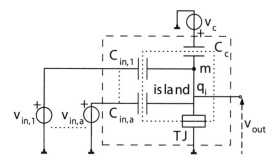

Fig. 13.3 Electron-box with a extra input capacitors.

(remember, the background charge and the initial charge are assumed to be zero. When the initial charge and background charge are not zero the control voltage v_C should be tuned until the effect of those charges is cancelled.) The critical voltage of junction TJ is equal to

$$v_j^{cr} = \frac{e}{2C_\Sigma};$$ (13.3)

an electron tunnels through tunnel junction as soon as the critical voltage is reached.

-A binary programmable building block

For the use of digital blocks it is of importance to define the logical levels. In this part high input and low input voltages are defined as, respectively, v_{ih}, v_{il}. For the output voltage also a high and a low output voltage is defined, respectively, v_{oh}, v_{ol}. The output voltage is the voltage over the junction. Further more is defined:

$$v_{il} = -v_{ih}$$ (13.4)

$$v_{ol} = -v_{oh}$$ (13.5)

In general, when the output level is lower than the input level we need amplification to be able to connect to the next stage. And, in case the output level is higher than the input level we need to attenuate the output. Without tunneling the output signal of an electron-box will always attenuate the input signal. To have some fan-out we need amplification. This can be obtained by using the bias voltage v_C to bias the junction, close to its tunnel condition. Now only small input voltages are necessary to let an electron tunnel, and by doing so a higher signal at the output can be reached than the original input signal. Depending on the bias value, an operating window is chosen such that different logic functions can be

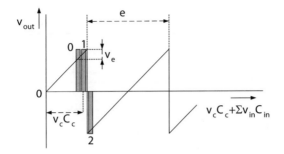

Fig. 13.4 The input-output relation of an electron-box biased to create a two input NAND function.

obtained. By changing the value of the bias voltage a NAND, NOR or inverter gate can be selected, we have a "programmable" logic gate.

-NAND-gate

To implement a NAND function, the operating window is placed in such a way that the output voltage is always high, except when the two inputs are high. In figure Fig. 13.4 the input-output relation of a two input electron-box is shown, in which the number indicates the number of inputs that is high (see also table 13.1). If both inputs are high (for a two input NAND) they give a low output voltage.

Between '1' and '2' an electron should tunnel (see the jump in voltage in figure Fig. 13.4). For the NAND we need to push an electron from the island ($q_i \rightarrow q_i - e$) when both of the inputs are high ($\sum v_{in} = 2v_{ih}$), that is, the effect of all voltages including the bias voltage must be higher than the critical voltage. Using equation 13.2 and equation 13.3 we obtain:

$$2v_{ih}C_{in} + v_CC_C > \frac{e}{2} \tag{13.6}$$

And the electron should not leave the island when one of the inputs is low ($\sum v_{in} = 0$), that is, the effect of all voltages including the bias voltage must be lower than the critical value:

Table 13.1 The 2-input NAND function, with $v_{il} = -v_{ih}$

v_{in1}	v_{in2}	$\sum v_{in}$	number of "high" inputs	v_{out}
v_{il}	v_{il}	$2v_{il}$	0	v_{ol}
v_{il}	v_{ih}	0	1	v_{ol}
v_{ih}	v_{il}	0	1	v_{ol}
v_{ih}	v_{ih}	$2v_{ih}$	2	v_{oh}

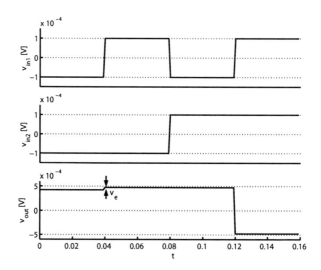

Fig. 13.5   Simulation results of the electron-box structure biased as a NAND, with $C_j = C_C = C_{in} = 40aF$, $v_C = 1.9mV$. The error voltage is also visible (at $t = 0.04s$). This error voltage is equal to $v_e = 50\mu V$.

$$v_C C_C < \frac{e}{2}. \tag{13.7}$$

Because we have chosen for positive and negative binary signals, the window '1' is true for $\sum v_{in} = 0$, and hence the position of this window is completely determined by the bias condition. The DC operating point should be within window '1'. The output voltage should change as little as possible in case one input is low and in case both inputs are low. From this we obtain the condition:

$$2v_{il} C_{in} \ll e \tag{13.8}$$

When both inputs are low a difference in output voltage will arise compared to the situation where only one input is low, due to the slope of C_{in}/C_Σ in the input-output relation (see also figure Fig. 13.4). Let v_e be the maximal difference in output voltage, then we can write:

$$v_e = 2|v_{il}|\frac{C_{in}}{C_\Sigma} \tag{13.9}$$

This error voltage should be as small as possible. The output drop should be as large as possible if compared with the error:

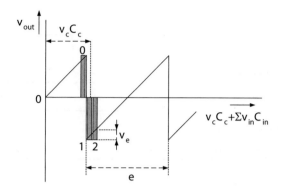

Fig. 13.6 The input-output relation of an electron-box biased to create a two input NOR function.

$$2|v_{il}|\frac{C_{in}}{C_\Sigma} \ll \frac{e}{C_\Sigma} \tag{13.10}$$

Which is in agreement with equation 13.8. Figure Fig. 13.5 shows simulation results for the electron-box biased as a NAND-gate.

-NOR-gate

To implement a NOR function, the operating window should be biased in such a way that the output voltage is always low, except for the case that both inputs are low. Figure Fig. 13.6 shows the input-output relation biased as a NOR (see table 13.2).

The equations to describe all variables are:

$$v_C C_C > \frac{e}{2} \tag{13.11}$$

and

$$2v_{il}C_{in} + v_C C_C < \frac{e}{2} \tag{13.12}$$

Table 13.2   The 2-input NOR function, with $v_{il} = -v_{ih}$.

v_{in1}	v_{in2}	$\sum v_{in}$	number of "high" inputs	v_{out}
v_{il}	v_{il}	$2v_{il}$	0	v_{ol}
v_{il}	v_{ih}	0	1	v_{oh}
v_{ih}	v_{il}	0	1	v_{oh}
v_{ih}	v_{ih}	$2v_{ih}$	2	v_{oh}

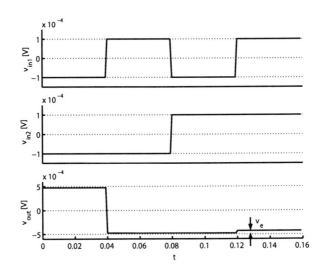

Fig. 13.7 Simulation results of the electron-box structure biased as a NOR, with $C_j = C_C = C_{in} = 40aF$, $v_C = 2.1mV$. In this figure the error voltage v_e is visible (beginning at $t = 0.12s$). This error is equal to $v_e = 50\mu V$.

An error is obtained between the situation of $\sum v_{in} = 1$ and $\sum v_{in} = 2$, due to a slope in the input-output characteristic of the electron-box. Let v_e be this error voltage, which is equal to

$$v_e = 2v_{ih}\frac{C_{in}}{C_\Sigma} \tag{13.13}$$

This error should be small compared to the voltage drop due to a tunnel event:

$$2v_{ih}\frac{C_{in}}{C_\Sigma} \ll \frac{e}{C_\Sigma} \tag{13.14}$$

hence

$$2v_{ih}C_{in} \ll e \tag{13.15}$$

In figure Fig. 13.7 the simulation results are shown for the electron-box biased as a NOR.

 - *The inverter*

It is also possible to make the inverter from a NAND or a NOR by respectively making one input high or low. But a direct implementation only uses one input. The equations describing the variables are:

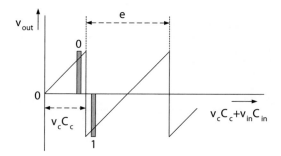

Fig. 13.8 The input-output relation of an electron-box biased to create an inverter function.

$$v_{ih}C_{in} + v_C C_C > \frac{e}{2}, \tag{13.16}$$

$$v_{il}C_{in} + v_C C_C < \frac{e}{2}, \tag{13.17}$$

and

$$v_C = \frac{e}{2C_C} \tag{13.18}$$

Shown are the input output relation and the simulation, respectively.

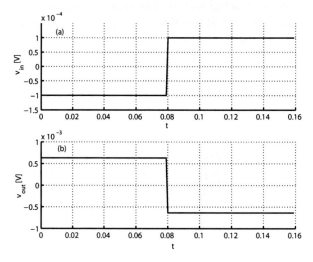

Fig. 13.9 The simulation results of the electron-box structure biased as an inverter, with $C_j = C_C = C_{in} = 40aF$, and $v_C = 2mV$.

Table 13.3   The values of the bias voltages for three different digital functions.

Function	v_C
NAND	$1.9mV$
NOR	$2.1mV$
Inverter	$2.0mV$

-Programmable logic

The logical functions implemented with the electron box depend on the applied bias voltage. By changing the bias voltage by a controller the total input-output relation can be controlled. For the three digital functions, NAND, NOR, and inverter described above, the necessary bias conditions are shown in table 13.3.

- Threshold function

The threshold function checks wether the number of "high" inputs is larger than a certain threshold value. The input output relation is shown in figure Fig. 13.10.

-EXOR function

More logic functions are possible, for example the important EXOR function. The obtained logic family is large enough to construct every Boolean function (for example, a full adder). The input-output relation of the electron box creating the EXOR function and the simulation results are shown in figure Fig. 13.11 and figure Fig. 13.12, respectively.

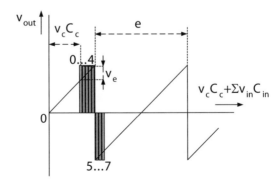

Fig. 13.10   The input-output relation of an electron-box with the chosen multiple operating windows to create a threshold function of < 5.

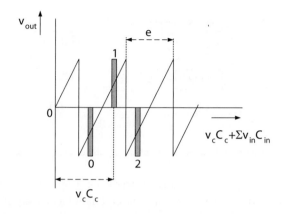

Fig. 13.11 The input-output relation of an electron-box biased to create an EXOR function.

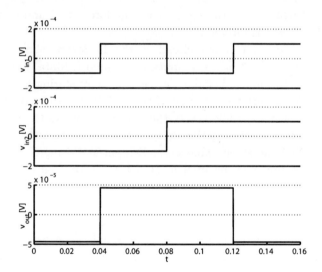

Fig. 13.12 The simulation results of the electron-box structure biased as a two input EXOR, with $C_j = C_C = 40aF$, $v_C = 5mV$, $C_{in} = 400aF$.

13.1.2 *Memory elements*

Based on the Boolean logic gates, memory elements can be design in the same way they are design in digital logic circuits. A pair of cross-coupled inverters can be used to hold two stable states. To force the circuit into any particular state a **R-S latch** can be used. The R-S latch consists of two

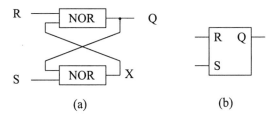

Fig. 13.13 R-S latch based on Boolean gates: (a) the circuit; (b) shorthand diagram.

cross-coupled NOR gates that form a feedback loop, see figure Fig. 13.13. When S= 0 and R= 0, the R-S latch circuit is logically equivalent to two cross coupled inverters, it has two stable states and can act as a memory. If S or R are changed the value stored by this circuit can be altered. If S becomes 1 while R= 0, the signal X will be forced to 0 and Q will be 1; if R becomes 1 while S= 0, the signal X will be forced to 1 and Q will be 0. The input S=R= 1 is forbidden.

A simple *clocked* memory circuit is the **D latch**, shown in figure Fig. 13.14. The D latch has two inputs, designated D (data) and G (gate); like the R-S latch, it is a bistable device capable of storing a single bit of information for an arbitrary period of time. So long as G= 0, the Q output remains constant; while G= 0, the output Q follows the value at the input D after a short propagation delay. The state of the latch can thus be changed by applying a new value to the D input and rising G momentarily. The AND function may be implemented by a NAND gate follow by an inverter.

-Single island memory
The electron box structure can be used as a dynamic memory cell (DRAM) that stores one or more electrons on the island. The island is for this purpose is in close proximity to a narrow conducting channel (for example,

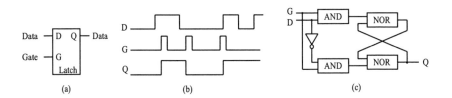

Fig. 13.14 D latch based on Boolean gates: (a) shorthand diagram, (b) operation, and (c) implementation.

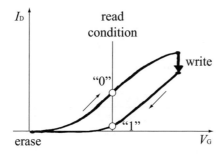

Fig. 13.15 Schematic voltage-current curve of the Yano memory.

that of a MOS transistor or a FET). Memory operation is achieved as follows. Electrons tunnel that tunnel to the island, due to an appropriate voltage on the gate electrode, change the threshold voltage of the narrow channel. Reducing the voltage, now, not immediately results in a tunnel event whereby the electron returns. The is this way obtained hysteresis can be exploited for a memory cell. One such a memory design is the so-called Yano memory [Yano *et al.* (1994)]; figure Fig. 13.15 shows the *vi*-characteristic of such a memory schematically.

13.2 Analog Functionality

Based on the SET junctions analog applications and devices have been proposed. For example, based on the oscillatory behavior of the tunneling junction (oscillation of the voltage across the junction) current controlled oscillators have been proposed. Some less obvious analog devices are discussed.

13.2.1 *Voltage controlled variable capacitor*

Earlier we saw in chapter 10 that the charge on the capacitor of an electron box structure can be expressed as:

$$q_{\rm C} = \frac{C_{\rm C}C_{\rm TJ}}{C_{\rm C} + C_{\rm TJ}}v - \frac{C_{\rm C}}{C_{\rm C} + C_{\rm TJ}}q_{\rm i}. \tag{13.19}$$

If N electrons tunnel to the island $q_{\rm i} = N{\rm e}$, neglecting initial and background charges, then

$$q_{\mathrm{C}} = \frac{C_{\mathrm{C}}C_{\mathrm{TJ}}}{C_{\mathrm{C}}+C_{\mathrm{TJ}}}v - \frac{C_{\mathrm{C}}}{C_{\mathrm{C}}+C_{\mathrm{TJ}}}N e. \tag{13.20}$$

Now, a differential capacitance can be defined:

$$c(v) = \frac{\mathrm{d}q_{\mathrm{C}}}{\mathrm{d}v} = \frac{C_{\mathrm{C}}C_{\mathrm{TJ}}}{C_{\mathrm{C}}+C_{\mathrm{TJ}}} - \frac{eC_{\mathrm{C}}}{C_{\mathrm{C}}+C_{\mathrm{TJ}}}\frac{\mathrm{d}N}{\mathrm{d}v}. \tag{13.21}$$

The first term is the "normal" equivalent capacitance of the two capacitances in series, that of the capacitor and that of the junction. It doesn't depend on the applied voltage v. The second term, Δc, is related to the charging and discharging by tunneling towards or of the island. The differential capacitance is relevant when an alternating voltage of amplitude Δv is added to a constant dc voltage V, so that the total applied voltage oscillates between $V \pm \Delta v/2$. When the applied voltage is smaller than the critical voltage minus the maximum ac voltage, $v < v^{\mathrm{cr}} - \Delta v/2$, the number of additional electrons on the island is always zero and $c = C_{\mathrm{eq}}$. When the applied voltage is swept between $v < v^{\mathrm{cr}} - \Delta v/2$ and $v < v^{\mathrm{cr}} + \Delta v/2$ the island is charged and discharged during every oscillation. This increases the capacitance above C_{eq} by Δc with:

$$\frac{\Delta c}{C_{\mathrm{eq}}} = \frac{e}{C_{\mathrm{TJ}}}\frac{\mathrm{d}N}{\mathrm{d}v} \tag{13.22}$$

A smooth variation of the capacitance can be obtained by using an assembly of small islands [Carrey *et al.* (2004)], since all the islands do not have the same critical voltage. Below the critical voltage of the highest-capacitance island in the assembly, no island can be charged or discharged and the differential capacitance stays constant. Increasing the applied voltage increases the differential capacitance as more and more islands can be charged and discharged during each cycle. This increase reflects the distribution of critical voltages of the various tunnel junctions.

13.2.2 Charge detection

A quite straightforward application of the SET is sensing charge at the gate capacitor; it is a very sensitive electrometer. The working principle of the electrometer is based on the offset of the vi-characteristic of the SET transistor as the charge on the gate capacitor changes to values larger than zero. Small charges, between 0 and $e/2$, lead to measurable changes in the current through the SET transistor, see figure Fig. 13.16.

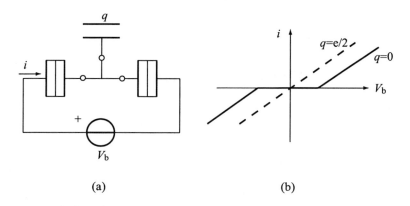

Fig. 13.16 (a) SET electrometer; (b) vi-characteristic of the electrometer for different values of sensed charge q.

13.2.3 *Electron pump in metrology*

If technology admits manufacturing of single-electron tunneling pumps producing currents with a relative uncertainty level of $1\text{x}10^{-8}$ then such an electron pump can be used as a quantum standard of current. As we saw in chapter 11 electrons can be "pumped" through many junctions by a proper clocking scheme on the gate voltage sources. In general one can write for the current in such a circuit:

$$i = nef, \tag{13.23}$$

where n is an integer telling how many electrons are pumped between two junctions and f is the pumping frequency. Having a quantum standard of current completes what is called "the metrological triangle", figure Fig. 13.17. The other two standards in this triangle are a voltage standard based on the superconductive Josephson effect and the standard of the resistance based on the quantum Hall effect. As you can the resistance value in the quantum Hall effect is the quantized resistance of chapter 5.

Problems and Exercises

13.1 Design a NAND logic gate in SET technology based on the circuit design methodology used in CMOS.

13.2 Design a NAND logic gate in SET technology based on the electron

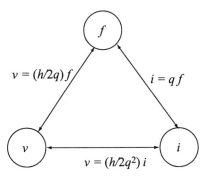

Fig. 13.17 Metrology triangle with the three material standards linked to the three quantum effects: Josephson effect, quantum Hall effect, and single-electron tunneling.

box.

13.3 Discuss the advantages and disadvantages of current CMOS logic compared to the nanoelectronic SET logic.

13.4 Design a EXNOR logic gate with electron-box logic, sketch the input-out relation.

13.5 Design a full adder based on electron-box logic.

13.6 Do a literature search on modern applications of single-electron circuits.

Epilogue

Readers, not new in the field of nanoelectronics and single-electron tunneling devices, might have noticed that some results obtained by modeling circuits with the impulse circuit model resemble the results obtained by modeling of those single-electron circuits with the so-called orthodox theory of single electronics [Likharev (1999)].

In the orthodox theory of single-electronics the tunnel junction is seen as a leaky capacitance and is modeled using a capacitance, C, and a tunnel resistance R_t that depends on the size of the junction and the barrier thickness. In this orthodox theory, two different types of analysis for explaining the behavior of circuits including SET junctions have been developed. The **local view** of Coulomb blockade, also called *tunnel junction in a high ohmic environment*, analyzes the tunnel junction as isolated from its environment, and only considers energies related to the tunnel junction. The **global view** of Coulomb blockade, also called *tunnel junction in a low ohmic environment*, analyzes the tunnel junction in cohesion with its environment, but does not take environmental resistances into account. An overall *change of energy*, called the free-energy, due to storage on junction capacitances and deliverance by sources, is taken into account. In general the local view is used when the resistance of the environment is much higher than R_Q, the global view is used when the resistance of the environment is much smaller than R_Q; R_Q is the resistance quantum: $R_Q = h/e^2 = 25.8\text{k}\Omega$.

The Coulomb blockade is explained as a property of an conducting island created by at least one tunnel junction and adjacent capacitors. When the island size becomes comparable with the De Broglie wavelength λ_F of the electrons inside the island, energy quantization becomes substantial—the De Broglie wavelength being $\lambda_F = h/\sqrt{2mE_F}$ for an electron at the Fermi-level E_F. In this case the charging effects can be best described by the electron addition energy E_a. The orthodox theory states that the energy necessary to tunnel onto an island is the Coulomb energy E_C, an assumption discussed critically in chapter 10. Now, if E_K is the quantum kinetic energy of the of the added electron due to the discreteness of the energy levels on the island, E_a may be well approximated by $E_a = E_C + E_K$. The orthodox theory considers *only islands of* such *'large' sizes* that E_a is dominated by E_C.

Furthermore, the theory makes the following three *major assumptions*: (1) the electron energy quantization inside the conducting island is ignored, (2) the non-vanishing duration of the tunnel time t_τ of electron tunneling through the barrier is assumed to be negligibly small, and (3) cotunneling is ignored. Globally, the *main result* of the orthodox theory can be formulated as follows: the tunneling of a single electron through a particular tunnel barrier is always a random event, with a certain rate Γ (probability per unit time) which depends solely on the reduction of the free energy of the *system* as a result of this tunneling.

If there is a *reduction of the system's free energy as a result of a single tunnel event*, tunneling will occur (if ΔF is the change in the free energy then in case of reduction: $\Delta F < 0$). If there is no reduction of free energy possible as a result of a tunnel event, then the system is in Coulomb blockade. The evaluations are done at T $\approx$ 0 K. The change of the free energy is defined as the change of the energy, w_{se}, stored on the junction capacitances minus the energy, w_s, delivered by the sources *during* a tunnel event. In terms of the change in free energy ΔF: If F_{before} and F_{after} are defined as the free energy before and after the tunnel event, respectively, and $w_{se|before}$ and $w_{se|after}$ as the stored energy on the junction capacitances before and after the tunnel event, then $\Delta F = (F_{after} - F_{before})$, $\Delta w_{se} = (w_{se|after} - w_{se|before})$, and:

$$\Delta F = \Delta w_{se} - w_s. \tag{14.24}$$

Tunneling is possible if:

$$\Delta F < 0, \tag{14.25}$$

and the criterium for the system to be in Coulomb blockade is:

$$\Delta F \geq 0. \tag{14.26}$$

From a circuit designer's point of view the reduction of free-energy has to be compensated by absorption of this energy by resistances (energy conservation). Because resistors are not taken into account (except for discriminating between local- and global view) the orthodox theory does not seem to be a good theory for designing real circuits. However, for calculating the critical voltage the orthodox theory use $\Delta F = 0$—that is, the criterion for energy conservation. One can show that the calculation of the critical voltage, in the global view, in the orthodox theory gives the same results as the calculation of the critical voltage in the impulse circuit model *provided that we consider only unbounded currents.*

As an example of this equivalence we consider ΔF during the tunneling event at junction TJ1 in the double tunnel junction structure. For calculating the critical voltage we start with the formulas of chapter 1 (equations 1.3 - 1.6). In *steady state* ($v(t) = v_s$), the expressions for the stored energy in the junctions TJ1 and TJ2 are (leaving out the time dependency):

$$w_{\text{se1}} = \frac{q_1^2}{2C_{\text{TJ1}}} = \frac{C_{\text{TJ1}}}{2(C_{\text{TJ1}} + C_{\text{TJ2}})^2}(C_{\text{TJ2}}^2 v_s^2 - 2C_{\text{TJ2}}v_s q_i + q_i^2) \quad (14.27)$$

$$w_{\text{se2}} = \frac{q_2^2}{2C_{\text{TJ2}}} = \frac{C_{\text{TJ2}}}{2(C_{\text{TJ1}} + C_{\text{TJ2}})^2}(C_{\text{TJ1}}^2 v_s^2 + 2C_{\text{TJ1}}v_s q_i + q_i^2). \quad (14.28)$$

Adding these two equations we find for the total stored energy w_{se}:

$$w_{\text{se}} = \frac{C_{\text{TJ1}}C_{\text{TJ2}}^2 + C_{\text{TJ1}}^2 C_{\text{TJ2}}}{2(C_{\text{TJ1}} + C_{\text{TJ2}})^2}v_s^2 + \frac{1}{2(C_{\text{TJ1}} + C_{\text{TJ2}})}q_i^2, \quad (14.29)$$

and for the change in stored energy during the tunnel event:

$$\Delta w_{\text{se}} = \frac{1}{2(C_{\text{TJ1}} + C_{\text{TJ2}})}(q_{i|\text{after}}^2 - q_{i|\text{before}}^2). \quad (14.30)$$

If we now define δq_s as the charge that is transported through the source during the tunnel event then:

$$w_s = \delta q_s v_s, \quad (14.31)$$

and we can calculate the condition for tunneling:

$$\Delta F < 0, \quad (14.32)$$

that is:

$$\Delta w_{\text{se}} - w_s < 0, \quad (14.33)$$

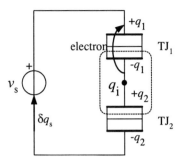

Fig. 14.18   The global view of the orthodox theory of single electronics: an electron tunnels through junction TJ1 of the double-junction structure. Due to $T \approx 0K$ the electron can only tunnel to the positive side of the junction.

or

$$w_s - \Delta w_{se} > 0. \tag{14.34}$$

We will do the calculation for zero initial island charge. And consider the tunneling through junction TJ1, see figure Fig. 14.18. When the electron tunnels from the island, the island charge becomes e (e $= 1.60 \times 10^{-19}$ C), that is: $q_{i|after} = $ e. To find the value of the current through the source, δq_s, consider the change in the charge on the negative side of junction TJ2:

$$\delta q_s = (-q_{2|before}) - (-q_{2|after}) = \frac{C_{TJ2}}{C_{TJ1} + C_{TJ2}}(q_{i|after} - q_{i|before}). \tag{14.35}$$

If we, now, fill in the values of $q_{i|before}$ and $q_{i|after}$, we obtain for the tunnel condition (equation 14.34) using equations 14.30 and 14.31:

$$\frac{C_{TJ2}}{C_{TJ1} + C_{TJ2}}ev_s - \frac{e^2}{2(C_{TJ1} + C_{TJ2})} > 0. \tag{14.36}$$

And we obtain as a condition for tunneling:

$$v_s > \frac{e}{2C_{TJ2}}. \tag{14.37}$$

We could have done the same calculation in the case that the electron tunnels towards the island through junction TJ2 and had the result:

$$v_s > \frac{e}{2C_{TJ1}}, \tag{14.38}$$

thus we can formulate that tunneling will occur as soon as:

$$v_{\mathrm{s}} > \frac{\mathrm{e}}{2C_{\max}}, \qquad \text{that is} \qquad v_{\mathrm{s}}^{\mathrm{cr}} = \frac{\mathrm{e}}{2C_{\max}}. \tag{14.39}$$

We notice that the result is equivalent with the result in chapter 11, equation 11.10—that is, the result is equivalent with the calculation in the impulse circuit model for unbounded currents.

This result is *not* a surprise. To see this we equate equation 14.36 to zero:

$$\frac{C_{\mathrm{TJ2}}}{C_{\mathrm{TJ1}} + C_{\mathrm{TJ2}}}\mathrm{e}v_{\mathrm{s}} - \frac{\mathrm{e}^2}{2(C_{\mathrm{TJ1}} + C_{\mathrm{TJ2}})} = 0. \tag{14.40}$$

Looking at the voltages across the junction (junction TJ1) we see that, in particular for this case:

$$v_{\mathrm{before}} = \frac{C_{\mathrm{TJ2}}}{C_{\mathrm{TJ1}} + C_{\mathrm{TJ2}}}v_{\mathrm{s}},$$

$$v_{\mathrm{after}} = \frac{C_{\mathrm{TJ2}}}{C_{\mathrm{TJ1}} + C_{\mathrm{TJ2}}}v_{\mathrm{s}} - \frac{\mathrm{e}}{C_{\mathrm{TJ1}} + C_{\mathrm{TJ2}}}, \tag{14.41}$$

so that:

$$\frac{\mathrm{e}}{2}(v_{\mathrm{before}} + v_{\mathrm{after}}) = \frac{C_{\mathrm{TJ2}}}{C_{\mathrm{TJ1}} + C_{\mathrm{TJ2}}}\mathrm{e}v_{\mathrm{s}} - \frac{\mathrm{e}^2}{2(C_{\mathrm{TJ1}} + C_{\mathrm{TJ2}})}. \tag{14.42}$$

This result is the same as in equation 14.40. We see that, in this case, the equation for the critical voltage can be written as a function of the *sum* of the voltages across the first junction before and after the tunnel event:

$$0 = -\frac{\mathrm{e}}{2}(v_{\mathrm{before}} + v_{\mathrm{after}}), \tag{14.43}$$

or

$$v_{\mathrm{after}} = -v_{\mathrm{before}}. \tag{14.44}$$

And we recognize the tunnel condition of the impulse circuit model based on the extended hot-electron model.

Bibliography

Alonso, M. and Finn, E. (1967). *Fundamental University Physics* (Addison-Wesley Publ. Co.).

Averin, D. and Likharev, K. (1986). Coulomb blockade of single-electron tunneling, and coherent oscillations in small tunnel junctions, *Journal of low temperature physics* **62**, 3/4, pp. 345–373.

Bylander, J., Duty, T. and Delsing, P. (2005). Current measurement by realtime counting of single electrons, *Nature* **434**, pp. 361–364.

Carrey, J., Seneor, P., Lidgi, N., Jaffres, H., Dau, F. N. V., Fert, A., Friederich, A., Montaigne, F. and Vaures, A. (2004). Capacitance variation of an assembly of clusters in the coulomb blockade regime, *Journal of Applied Physics* .

Chua, L., Desoer, C. and Kuh, E. (1987). *Linear and Nonlinear Circuits* (McGraw-Hill).

Csurgay, A. (2007). On circuit models for quantum-classical networks, *International Journal of Circuit Theory and Applications* **35**, pp. 471–484.

Csurgay, A. and Porod, W. (2004). Special issue on nanoelectronic circuits, *International Journal of Circuit Theory and Applications* **32**, pp. 275–446.

Csurgay, A. and Porod, W. (2007). Special issues on nanoelectronic circuits, *International Journal of Circuit Theory and Applications* **35**, pp. 211–390.

Davenport, W. and Root, W. (1958). *An Introduction to the theory of Random Signals and Noise*, chap. 7 Shot Noise (McGraw-Hill Book Company, Inc.), pp. 112–144.

Davies, J. (1998). *The Physics of Low-Dimensional Semiconductors* (Cambridge University Press).

Davis, A. (1994). A unified theory of lumped circuits and differential systems based on heaviside operators and causality, *IEEE Transactions on Circuits and Systems-I* **41**, pp. 712–727.

Davis, A. (1998). *Linear circuit analysis* (PWS Publishing Company, ISBN 0-534-95095-7, Boston).

Devoret, M., D.Esteve, Grabert, H., Ingold, G., Pothier, H. and C.Urbina (1990). Effect of the electromagnetic environment on the coulomb blockade in ultrasmall tunnel junctions, *Physical Review Letters* **64**, 15, pp. 1824–1827.

Dragoman, M. and Dragoman, D. (2006). *Nanoelectronics: Principles and Devices*

(Artech House, Inc.).

Esaki, L. (1958). New phenomenon in narrow germanium p-n junctions, *Phys. Rev.* **109**, pp. 603–604.

Feynman, R., Leighton, R. and Sands, M. (1964). *The Feynman Lectures on Physics* (Addison-Wesley Publ. Co.).

Fromhold, J. A. T. (1981). *Quantum mechanics for applied physics and engineering* (Dover Publications Inc., New York).

Fulton, T. and Dolan, G. (1987). Observation of single-electron charging effects in small tunnel junctions, *Physical review letters* **59**, 1, pp. 109–112.

Geerligs, L., Anderegg, V., Holweg, P., Mooij, J., Pothier, H., Esteve, D., Urbina, C. and Devoret, M. (1990). Frequency-locked turnstile device for single electrons, *Physical review letters* **64**, 22, pp. 2691–2694.

Giaever, I. (1960). Electron tunneling between two superconductors, *Physical Review Letters* **5**, 10, pp. 464–466.

Giaever, I. (1969). Metal-insulator-metal tunneling, in *Tunneling Phenomena in Solids*, chap. 3 (Plenum Press, Edited by E. Burstein and Lundqvist, New York), pp. 19–30.

Giaever, I. and Zeller, H. (1968). Superconductivity of small tin particles measured by tunneling, *Physical Review Letters* **20**, 26, pp. 1504–1507.

Goossens, M., Verhoeven, C. and van Roermund, A. (1995). Conceps for ultra-low power and very-high-density single-electron neural networks, in *NOLTA '95*, Vol. 7B-7 (International Symposium on Nonlinear Theory and its Applications, Las Vegas, U.S.A.), pp. 679–682.

Goser, K., Glosekotter, P. and Dienstuhl, J. (2003). *Nanoelectronics and Nanosystems. From transistors to molecular and quantum devices* (Springer).

Haar, R. van de, and Hoekstra, J. (2003). *Simulation of a neural node using SET technology, Lecture Notes in Computer Science 2606*, Vol. Evolvable Systems: From Biology to Hardware (Springer-Verlag), pp. 377–386.

Hanson, G. W. (2008). *Fundamentals of Nanoelectronics* (Pearson Prentice Hall (International Edition)).

Heerkens, C. Th. H., Kamerbeek, M. J., van Dorp W. F., Hagen, C. W., and Hoekstra, J. (2008). Electron Beam Induced Deposited Etch Masks, *Microelectronics Engineering* (Elsevier).

Heij, C., Hadley, P. and Mooij, J. (2001). Single-electron inverter inverter, *Applied Physics Letters* **78**, p. 1140.

Henderson, A. (1971). Continuity of inductor currents and capacitor voltages in linear networks containing switches, *International Journal of Electronics* .

Hoekstra, J. (1987). Junction charge-coupled devices for bit-level systolic arrays, *IEE Proceedings Electronic Circuits and Systems* **134**, G.

Hoekstra, J. (2001). On the origin of energy loss in single-electron tunneling devices, in *Proceedings of the ninth workshop on Nonlinear Dynamics of Electronic Systems*, NDES (ISBN 90-407-2194-7, Delft, The Netherlands), pp. 117–120.

Hoekstra, J. (2004). On the impulse circuit model for the single-electron tunnelling junction, *International Journal of Circuit Theory and Applications* **32**, pp. 303–321.

Hoekstra, J. (2007a). On circuit theories for single-electron tunnelling devices, *IEEE Trans. on Circuits and Systems 1: Regular Papers* **54**, pp. 2353–2360.

Hoekstra, J. (2007b). Towards a circuit theory for metallic single-electron tunnelling devices, *Intenational Journal of Circuit Theory and Applications* **35**, pp. 213–238.

Hoop, A. J. de (1975). *Theorie van het elektromagnetische veld* (Nijgh-Wolters-Noordhoff).

Ingold, G. and Nazarov, Y. (1992). Charge tunneling rates in ultrasmall junctions, in *Single Charge Tunneling, Coulomb Blockade Phenomena in Nanostructures, NATO ASI series B*, Vol. 294, physics edn., chap. 2 (Plenum Press, New York), pp. 21–107.

Kasper, E. and Paul, D. (2005). *Silicon Quantum Integrated Circuits*, Nanoscience and Technology (Springer).

Kauppinen, J. and Pekola, J. (1996). Charging in solitary, voltage biased tunnel junctions, *Physical Review Letters* **77**, 18, pp. 3889–3892.

Kirihara, M. and Taniguchi, K. (1997). A single electron neuron device, *Japanese Journal of Applied Physics* **36**, 6B, pp. 4172–4175.

Klauw, C. van der(1986). A full adder using junction charge-coupled logic, *IEEE Journal of Solid-State Circuits* **SC-21**, 4, pp. 584–587.

Klunder, R. and Hoekstra, J. (2001). Energy conservation in a circuit with single electron tunnel junctions, in *The IEEE international Symposium on Circuits and Systems* (ISCAS, Sydney, Australia), pp. I–591 – I–594.

Klunder, R. and Hoekstra, J. (2004). The synthesis of an exor function by using modulo functions implemented by set circuits, in *Scientific Computing in Electrical Engineering*, h.a. schilders et al. (eds.) edn., ECMI (Springer), pp. 273–280.

Lambe, J. and Jaklevic, R. (1969). Charge-quantization studies using a tunnel capacitor, *Physical Review Letters* **22**, 25, pp. 1371–1375.

Lent, C., Tougaw, P., Porod, W. and Bernstein, G. (1993). Quantum cellular automata, *Nanotechnology* **4**, pp. 49–57.

Likharev, K. (1999). Single-electron devices and their applications, in *Proceedings of the IEEE*, Vol. 87 (IEEE), pp. 606–632.

Lindmayer, J. and Wrigley, C. (1965). *Fundamentals of Semiconductor Devices* (D. van Nostrand Company).

Lotkhov, S., Bogoslovsky, S., Zorin, A. and Niemeyer, J. (2001). Operation of a three-junction single-electron pump with on-chip resistors, *Applied Physics Letters* **78**, 7, pp. 946–948.

Mizutani, U. (2001). *Introduction to the Electron Theory of Metals* (Cambridge University Press).

Pauli, W. (1973). *Electrodynamics* (Dover).

Roermund, A. van, and Hoekstra, J. (2000). Design philosophy for nanoelectronic systems, from sets to neural nets, *International Journal of Circuit Theory and Applications* **28**, 6, pp. 563–584.

Rosenblatt, F. (1962). *Principles of Neurodynamics: Perceptrons and the Theory of Brain Mechanisms* (Spartan Books).

Spice is developed at University of California (at Berkeley), (1975). *SPICE is a general-purpose circuit simulation program* .

Sze, S. (1981). Tunnel devices, in *Physics of secmiconductor devices*, 2nd edn., chap. 9 (Wiley), pp. 513–565.

Sze, S. and Ng, K. K. (2007). *Physics of Semiconductor Devices - 3rd Edition* (Wiley).

Thornber, K., McGill, T. and Mead, C. (1967). The tunneling time of an electron, *Journal of Applied Physics* **38**, 5, pp. 2384–2385.

Wasshuber, C., Kosina, H. and Selberherr, S. (1997). Simon - a simulator for single electron tunnel devices and circuits, *IEEE transactions on computer aided design of integrated circuits and systems* **16**, 9, pp. 937–944.

Wernersson, L.-A., Kabeer, S., Zela, V., Lind, E., Zhang, J., Seifert, W., Kosel, T. and Seabaugh, A. (2005). A combined chemical vapor deposition and rapid thermal diffusion process for sige esaki diodes by ultra-shallow junction formation, *IEEE Transaction on Nanotechnology* **4**, pp. 594–598.

Yano, K., Ishii, T., Hashimoto, T., Kobayashi, T., Murai, F. and Seki, K. (1994). Room-temperature single-electron memory, *IEEE transactions on electron devices* **41**, 9, pp. 1628–1638.

Zimmerman, T., Allen, R. and Jacobs, R. (1977). Digital charge-coupled logic (dccl), *IEEE Journal of Solid-State Circuits* **SC-21**, 5, pp. 473–485.

Index